Guidelines for Analysis and Description of Soil and Regolith Thin Sections

Second Edition

Georges Stoops

Emeritus Professor, Department of Geology, Faculty of Sciences, Ghent University, Ghent, Belgium.

partially based upon the

"Handbook for Soil Thin Section Description" by Bullock, P., Fedoroff, N., Jongerius, A., Stoops, G., Tursina, T. and Babel, U.

Soil Science Society of America

WILEY

Copublication by © American Society of Agronomy, Inc., Crop Science Society of America, Inc., and Soil Science Society of America, Inc. and John Wiley & Sons, Inc.

Limit of Liability/Disclaimer of Warranty
While the publisher and author have used their best efforts in preparing this book, they make no representations or warranties with respect to the accuracy or completeness of the contents of this book and specifically disclaim any implied warranties or merchantability of fitness for a particular purpose. No warranty may be created or extended by sales representatives or written sales materials. The publisher is not providing legal, medical, or other professional services. Any reference herein to any specific commercial products, procedures, or services by trade name, trademark, manufacturer, or otherwise does not constitute or imply endorsement, recommendation, or favored status by the ASA, CSSA and SSSA. The views and opinions of the author(s) expressed in this publication do not necessarily state or reflect those of ASA, CSSA and SSSA, and they shall not be used to advertise or endorse any product.

Editorial Correspondence:
American Society of Agronomy, Inc.
Crop Science Society of America, Inc.
Soil Science Society of America, Inc.
5585 Guilford Road, Madison, WI 53711-58011, USA

agronomy.org
crops.org
soils.org

Registered Offices:
John Wiley & Sons, Inc., 111 River Street, Hoboken, NJ 07030, USA

For details of our global editorial offices, customer services, and more information about Wiley products, visit us at www.wiley.com.

Wiley also publishes its books in a variety of electronic formats and by print-on-demand. Some content that appears in standard print versions of this book may not be available in other formats.

Library of Congress Cataloging-in-Publication Data

Names: Stoops, Georges, author.
Title: Guidelines for analysis and description of soil and regolith thin sections /
  Georges Stoops, University of Ghent (Belgium), Faculty of Sciences,
  Department of Geology, Research Unit Mineralogy and Petrography.
Other titles: Guidelines for analysis and description of soil and regolith
  thin sections
Description: Second edition. | Hoboken, NJ : Wiley-ACSESS, [2021] | Series:
  ASA, CSSA, and SSSA books | Revised edition of: Guidelines for analysis
  and description of soil and regolith thin sections. Georges Stoops,
  Michael J. Vepraskas. 2003. | Includes bibliographical references and
  index.
Identifiers: LCCN 2020030573 | ISBN 9780891189756 (paperback)
Subjects: LCSH: Soil micromorphology. | Regolith.
Classification: LCC S593.2 .S76 2021 | DDC 631.4/3—dc23
LC record available at https://lccn.loc.gov/2020030573

doi:10.2136/guidelinesforanalysis2

Cover Design: Wiley
Cover Image: © Georges Stoops
10 9 8 7 6 5 4 3 2 1

To my wife Marthe for her patience during the many hours, days, weeks of writing, and for her continuous moral support. As a souvenir to the many "holidays" in the mountains and seaside during which piece by piece, year by year most of these notes were prepared.

Table of Contents

About the Second Edition

Since its publication, fifteen years ago, the *"Guidelines for Analysis and Description of Soil and Regolith Thin Sections"* are internationally considered as a standard for micromorphological studies, as a follow-up of the famous *"Handbook for Thin Section Description"*, developed by the International Working Group on Soil Micromorphology of the International Society of Soil Science (Bullock et al., 1985). As the first edition is out of print since a few years, and second-hand copies only available at exaggerated prices, that students surely cannot afford, I took the initiative to prepare a second, updated version.

Since the publication of the first edition, much progress was made in the interpretation of micromorphological features, especially in the fields of archaeology and paleopedology. However, new publications on methods, theoretical concepts and terminology are very rare. The study of many papers applying the concepts of the Guidelines, and especially by refereeing many manuscripts, learned me which concepts and definitions were not clear or insufficiently explained. Also the discussion with students during several intensive courses on micromorphology helped me to discover what had to be remediated.

In this second edition, the text is updated, not only with new references, but also with some older that were overlooked before. Some chapters are rearranged, part of the appendixes integrated as tables in the corresponding chapters, other deleted. A new appendix, containing the translation of 220 terms in 19 languages is added.

I want to extend my thanks for useful comments and suggesting to several colleagues, especially Dr. D. Itkin (Ben-Gurion University of the Negev), Dr. Vera Marcelino (Ghent University, Belgium) and Dr. Florias Mees (Africamuseum and Ghent University, Belgium), and for the interesting comments of many referees, especially of Dr. M. Gerasimova (Moscow Lomonosov University), Dr. R. Heck (University of Guelph), Dr. F. Khormali (Gorgan University of Agriculture and Natural Resources), Dr. P. Kühn (University of Tübingen), Dr. V. Marcelino (Ghent University), Dr. H. Morrás (University of Buenos Aires), Dr. R. Poch (University of Lleida), L. Trombino (University of Milano), E. Van Ranst (Ghent University),

Dr. M. Vepraskas (North Carolina State University), Dr. E. Verrecchia (University of Lausanne), and anonymous referees.

My gratitude goes also to Prof. Dr. J. De Grave, head of the Research Unit Mineralogy and Petrology of the Department of Geology of the Ghent University (Belgium) for giving me the opportunity to make use of the infrastructure and thin section collections.

GEORGES STOOPS

Acknowledgements to the First Edition

While preparing the texts and illustrations, the author has often recalled all the individuals which have contributed in one way or another over many years to the final production of this book through their works, discussions, support, or advise.

My thanks go in the first place to the members of the former "Working Group on Soil Micromorphology", who regularly assisted at the meetings, and soon became dear friends. The long, sometimes seemingly endless discussions we had on concepts and their formulation were an excellent school where I learned the value of clear, unambiguously expressed ideas and where I became aware that cultural diversity, as small as it might seem within Western Europe, can lead to completely different approaches of scientific subjects, particularly when abstract ideas are involved. Discussions by Dr. H.J. Altemüller and the late Dr. A. Jongerius contributed much to the understanding of the micromorphological concepts needed, but every member of the Working Group had input, not only in a scientific way, but also in a human one. To say it with the words of the popular German singer Reinhard Mey: "Denn eigentlich ging keiner fort: in einer Geste, einem Wort, in irgend einer Redensart lebt Ihr in meiner Gegenwart" (Then in fact none [of my former friends] left. They all live close to me, in a gesture, a word, an expression).

I also want to acknowledge the contributions of numerous students to which I have taught micromorphology over the last 30 years, at the International Training Centre for Post-Graduate Soil Scientists, University of Gent, or as a visiting professor in Europe and overseas. Their remarks and questions helped me a lot to rephrase, reshape and complete definitions, subdivisions and comments. Looking to the *Handbook for Soil Thin Section Description* with the eyes of the student was very instructive, although, as a co-author, not always a very satisfactory experience.

Many individuals have contributed to the final result of this book. Mentioning every contribution would be impossible. Among those that sent in the past opinions on the *Handbook* I want to acknowledge especially Dr. H. Morras (INTA, Argentina) for his detailed and well-considered comments. Several colleagues improved the manuscript of the "Key to the ISSS Handbook for Soil Thin Section Description", which forms an essential part of this manual, determining the way in which several concepts were modified and interrelated. Amongst them I want to acknowledge the contributions of Dr. J. Arocena (UBC, Canada) and Dr. A. Ringrose-Voase (CSIRO, Australia). A great help to me were the comments of Dr. L. Drees (Texas A&M University, USA) on the first manuscript of this book, both with respect to the content and the redaction. Thanks go also to the members of the Editorial Committee who reviewed the manuscript. I really appreciate their comments and corrections. Following scientists were involved: Prof. A. Busacca (Washington State University), Dr. L. Drees (Texas A&M University), Prof. Dr. P. Goldberg (Boston University), Prof. Dr. R.W. Griffin (Prairie View A & M University, Texas), Dr. A. Jongmans (Wageningen Agricultural University, The Netherlands), Prof. Dr. D.L. Lindbo (North Carolina State University), Dr. W.D. Nettleton (USDA), Dr. F.E. Rhoton (USDA), Prof. Dr. H. Stolt (University of Rhode Island), Prof. Dr. L.T. West (The University of Georgia), Prof. Dr. L.P. Wilding (Texas A & M University) and Dr. M. Wilson (USDA).

Special thanks go to Dr. F. Mees (University of Gent, Belgium) for his comments on the first draft and his digitizing all the micrographs shown in the CD. Final formatting of the CD was done by Matthew Vepraskas (Virgin Tech. University) and Matthew Kirk (North Carolina State University). Thanks also to Mrs. Martine Bogaert (University of Gent, Belgium) for making the drawings.

Many colleagues kindly and spontaneously helped me with providing advises, information and illustrations. Not being able to thank everybody who gave some help, I want to mention especially following persons: Dr. F. Runge (Paderborn, Germany) and L. Vrijdaghs (Tervuren, Belgium) (phytoliths), Dr. J. Delvigne (Marseille, France) (weathering).

Although the collection of thin sections of the Laboratory for Mineralogy, Petrology and Micropedology of the Ghent University (Belgium) is very rich, some specific examples were missing, or not sufficiently didactic. The author is indebted to several friends and colleagues that send micrographs, especially Dr. J. van de Meer (University of Amsterdam, The Netherlands) (Micrographs 4.22 and 7.28), Dr. B. Van Vliet-Lanoë (University of Lille, France) (Micrographs 3.27 and 3.28), or gave the opportunity to make micrographs of their thin sections: Dr. J. Aguilar (University of Granada, Spain), Dr. C. Ampe and Dr. V. Marcelino (Ghent University, Belgium), Dr. P Kühn (Greifswald, Germany), Dr. R. Poch (University of Lleida, Spain), Dr. L. Trombino (University of Milano, Italy).

Thanks go also to Dr. C.V. Waine, publisher of the *Handbook for Soil Thin Section Description* for the permission to reproduce several figures.

Last, but not least, many thanks go to Dr. M. J. Vepraskas (North Carolina State University, USA), who urged me to write this book and who started and continued the timeconsuming and sometimes difficult administrative publishing procedures, including the editing. Without his help and continuous support this work would not have been realized.

GEORGES STOOPS

List of Abbreviations

BLF	blue-light fluorescence microscopy
CL	cathodoluminescence microscopy
CPL	circular polarized light
CT	X-ray computerized tomography
EDS	energy-dispersive spectroscopy
FTIR	Fourier-transformed infrared spectroscopy
OIL	oblique incident light
PPL	plane-polarized light
SEM	scanning electron microscopy
TDFI	transmitted dark field illumination
TEM	transmission electron microscopy
UVF	ultraviolet fluorescence microscopy
WDS	wavelength-dispersive spectroscopy
XPL	cross-polarized light
XPLλ	cross-polarized light and 1λ-retardation plate (gypsum compensator) inserted

1. Introduction

Precise descriptions of the features seen in soils or regoliths as examined under the microscope require a specific set of concepts and terms because the microscope reveals features that simply cannot be seen with the naked eye. Microscopic features can of course be described using common words, but this would lead to very tedious and lengthy descriptive texts that are time consuming both to write and to read and not always unambiguous. Moreover, it would be difficult to translate such descriptions without losing information or committing errors. By using a comprehensive terminology, descriptions would be not only shorter, but also easier to compare and to store in databases.

Terminology is in the first place a means of communication and, in the second place, a means of education- people more easily recognize objects, features, or situations for which they know a name. Features or combinations of features without a name are often not consciously observed! For instance, Inuits have many words for snow, while speakers of English have only one and can barely differentiate between wet and dry snow. Eunologues can distinguish and name many types of wines, based on the variety of grapes, fermentation and storing, whereas people not acquainted with this terminology can merely recognize red, white, and rosé wines.

To put an end to the proliferation of overlapping or contradictory concepts and terms in micromorphological publications, an international working group was created in 1969, under the auspices of the International Society of Soil Sciences, to establish a simple, comprehensive terminology for the description of soil thin sections. The result of this work was published in the Handbook for Soil Thin Section Description by P. Bullock, N. Fedoroff, A. Jongerius, G. Stoops, T. Tursina and U. Babel in 1985 (hereafter referred to as the Handbook). The book was highly appreciated by the micromorphological community, as it helped solve several problems of description inherent to the then existing systems. It became widely used, both for scientific research and as a teaching aid.

Since the early 1990s the Handbook had been out of print, but the original publisher was not interested in the publication of a second edition.

Guidelines for Analysis and Description of Soil and Regolith Thin Sections, Second Edition. Georges Stoops.
© 2021 Soil Science Society of America, Inc. Published 2021 by John Wiley & Sons, Inc.
doi:10.2136/guidelinesforanalysis2

Because of the demand for a new edition and to have the opportunity to amend several errors, contradictions and inconsistencies in the original text, I agreed to prepare a new revised text. The *Guidelines for Analysis and Description of Soil and Regolith Thin Sections* (hereafter referred to as the Guidelines) appeared in 2003. The text of this book was essentially based on the Handbook (Bullock et al., 1985), and on the author's own series of lecture notes and his experience in research and teaching at the International Training Centre for Post-Graduate Soil Scientists (Ghent University, Belgium) and during several intensive courses on micropedology in Europe and abroad. For some definitions and concepts, different approaches by other soil micromorphologists, which were discussed by Bullock et al. (1985), were not repeated in the Guidelines. Decisions then made, were adopted without arguments or references. In several places, however, definitions and schemes were discussed in more detail, as experience has shown that students are often puzzled why specific decisions were made.

Not all concepts of the Handbook were as user-friendly as intended by its authors. Especially in those cases where the distinction between features was partly based on common experience of the authors, some concepts were left unclear (Stoops and Tursina, 1992). Stoops (1998) suggested, therefore, the introduction of a key, which would probably not enhance the scientific level of the system much but would surely contribute to the use of unambiguous concepts and to a higher reproducibility of the descriptions, making it easier to store them in a database.

Almost 15 yr after its publication in 2003 the Guidelines was out of print, and a second, updated edition was urgently needed, as the system of concepts and terms became internationally the standard for micromorphological studies. In this second edition some concepts, giving rise to misunderstanding, are clarified and references to literature updated and extended. Almost no new ideas on description or concepts and terms were published in the last two decades. The concepts of the Guidelines were meanwhile also explained in two manuals: Loaiza et al., (2015) and Simões de Castro and Cooper (2019).

In the 1960s and the 1970s, micromorphology was often related to soil classification and/or related genetic studies. Since that time, application has gone beyond the bounds of traditional soil science as other disciplines discovered the utility of micromorphology. Other frequent users of micromorphology include: Quaternary geologists (e.g., Catt, 1989; Kemp, 1999; Cremaschi et al., 2018), sedimentologists (e.g., Zimmerle, 1991; van der Meer and Menzies, 2011; Menzies and van der Meer, 2018), weathering specialists (e.g., Nahon, 1991; Tardy, 1993; Delvigne, 1998), and especially archaeologists (e.g., Courty et al., 1989; Macphail et al., 1990; Davidson et al., 1992; Goldberg and Macphail, 2006; Macphail, 2008, 2014; Nicosia and Stoops, 2017; Goldberg and Aldeias, 2018; Macphail and Goldberg, 2018).

The objective of this book is to provide a system of analysis and description of soil and regolith materials as seen in thin sections. It is not intended as a manual of micropedology; topics such as sampling, thin section preparation, and interpretation of thin sections are therefore not discussed. Also, no attempt has been made to present proposals for higher levels of classification of microfabrics, as no sufficient agreement exists in the international micromorphological community on how to handle this problem.

In the past, many authors mixed the terminologies of Bullock et al. (1985) with those of Brewer (1964a and 1976), Brewer and Pawluk (1975) and others, without realizing the differences (e.g., differences in basic concepts) and especially without being aware of the false interpretations that might result. It is indeed scientifically incorrect to use a mixture of concepts and terms of different systems, which are not compatible. Is there any soil scientist that would accept a classification proposal for a soil profile, expressed in a mixture of U.S. Soil Taxonomy and WRB criteria and terms? Experience has shown that such a mixture of terms is dangerous and often leads to false statements.

To avoid confusion, some micromorphological concepts, definitions, and terms used by other systems are set off in separate explanatory paragraphs "Background", as complementary information to the reader, but not as a suggestion for its use as part of the proposed terminology. Where appropriate, concepts and terms are compared with those of other authors, without going into detail. The reader is referred to the original papers, or to Stoops and Eswaran (1986) or Jongerius and Rutherford (1979) for additional information. A complete glossary of existing micromorphological terms is beyond the scope of this textbook.

Terminology and/or classification reflect the state of the art in a given field of science and can therefore only be an approximation. The author is aware that this book is only a next approximation to a completer and more rational micropedological terminology.

2. Definitions and Historical Review

2.1 WHAT IS SOIL MICROMORPHOLOGY?

Soil micromorphology is a method of studying undisturbed soil and regolith samples with microscopic and ultramicroscopic techniques to identify their different constituents and to determine their mutual relations, in space and time. Its aim is to search for the processes responsible for the formation or transformation of soil in general, or of specific features, whether natural (e.g., clay skins, nodules) or artificial (e.g., irrigation crusts, plow pans), and their chronology. Consequently it is an important tool for investigations of soil genesis, classification, or management of soils and regoliths. The technique has also proven its usefulness in other domains, especially paleopedology and archeology.

A bibliometric study by Stoops (2014, 2018) shows that from 1950s onwards the number of micromorphological papers published increased, reaching a maximum of almost 700 during the period 1986 to 1990, decreasing slightly from then on. This decrease, explained mainly by the loss of interest in soil genesis and classification topics (due to shortage of funding) and the fact that discussions on new methods and concepts stabilized, was partly compensated by a gradual increase in the fields of paleopedology and archeology (see also Courty et al., 1989; Nicosia and Stoops, 2017; Adderley et al., 2018; Cremaschi et al., 2018; Fedoroff et al., 2018; Macphail and Goldberg, 2018).

Micromorphological investigations are based on the principles of (i) preservation of the fabric and structure, and (ii) functional investigation. Hence, the investigations should be performed on undisturbed and mostly naturally-oriented samples (in view of the characteristic vertical anisotropy of the soil), in contrast to the other analytical methods used in soil science. Chemical, physical, and mineralogical analyses usually require mixing, grinding, solubilization, or fractionation of representative soil samples and therefore yield average data. This is not the case for micromorphology, which often allows the examination of specific features in soils. According to the principle of functional investigation, all

Guidelines for Analysis and Description of Soil and Regolith Thin Sections, Second Edition. Georges Stoops.
© 2021 Soil Science Society of America, Inc. Published 2021 by John Wiley & Sons, Inc.
doi:10.2136/guidelinesforanalysis2

observations should be directed to the understanding of the function of each soil constituent or fabric within the soil as a whole.

Most microscopic observations of soil materials are made on thin sections. These are thin (30 μm) slices of a soil or regolith material that has been impregnated with plastic, glued to a glass slide, and then cut and polished to a thickness where the materials become translucent to light.

The research domain of micropedology covers all observations of undisturbed earthy samples under the microscope, including studies of thin sections, micromanipulations, microchemical and microphysical methods, and ultramicroscopic techniques. The best-developed and most popular part of micropedology is fabric analysis of thin sections, also called soil micromorphology, and its quantitative aspect, soil micromorphometry. Micromorphology is often used as a synonym for micropedology.

2.2 BRIEF HISTORICAL REVIEW

Observations made on soil materials using a hand lens, in either the field or laboratory, have probably been performed since the early beginning of soil science. Although the study of soil thin sections dates back to the beginning of the 20th century (Delage and Lagatu, 1904; Agafonoff, 1929, 1936a, 1936b; see also Stoops, 2009a, 2018), the first person to use magnifying instruments in a systematic way to study the soil was the Austrian scientist W.L. Kubiëna, considered therefore the "founding father of micropedology". He reported his first observations in some short papers in the early 1930s (Kubiëna, 1931), but his work received international recognition after the publication of his manual *Micropedology* in 1938, which was prepared during his stay as visiting professor in Iowa (Stoops, 2009b).

The scientific work of Kubiëna can be subdivided into two periods (Stoops and Eswaran 1986). In the first period, Kubiëna analyzed the fabric (internal organization) of the soil according to purely morphological criteria, using a morphoanalytical approach. The genetic interpretation of the morphology then followed. In his book *Micropedology*, Kubiëna (1938) defined different levels of fabric and gave an extensive description of the lowest level, the elementary fabric, as "the arrangement of the constituents of lowest order in soil in relation to each other", in other words the related distribution between stable coarse material (called skeleton grains, e.g., mineral grains, rock fragments) and the mobile fine material (called plasma, composed of colloids or clay). A terminology, partially consisting of newly coined terms, was introduced to name the different fabric types observed. In the second period, a morphogenetic approach prevailed, which means that specific combinations of soil features in soil thin sections were interpreted to explain the genesis of the soil material examined. Micropedology was at the base of Kubiëna's ideas on soil genesis and his new system of soil classification. These approaches were discussed comprehensively for the first time in his book *Entwicklungslehre des Bodens* (1948), and later in *The Soils of Europe* (1953) appearing simultaneously in Spanish, English and German, and in several papers in journals and proceedings. Most of his later ideas were published in his last book *Micromorphological Features in Soil Geography* (1970). The morphogenetic approach differs from the morphoanalytical one in that it is not limited to merely analyzing the fabric, but also directly involves a genetic interpretation of the observations. In fact, this morphogenetic approach of the microfabrics involves a genetic interpretation of the soil studied, right from the descriptive phase of study. No individual features are considered, but all characteristics as a whole are related to a specific soil type, after which the microfabric is named. Well-known terms are Braunlehm, Rotlehm, Braunerde and Roterde, which were presented in a hierarchic sequence, Braunlehm being at the origin of all other types. Also detailed micromorphological descriptions of humus types were given, from the terrestrial Mor to the subaquatic Anmoor. Kubiëna's approach to the soil microfabric was not purely analytical, but rather a personal view on specific aspects of soil formation, as seen under the microscope. A limitation of Kubiëna's system is that it was restricted to the soil types he described, and could not be used for soil materials with a similar fabric but a different genetic evolution. Moreover, his interpretations were generally not supported by other soil analyses (e.g., mineralogical, physical and/or chemical).

In the early 1960s, an expansion of micromorphology in different countries occurred, and it became clear that the morphogenetic approach of Kubiëna and his school was unsatisfactory. As a result, a new morphoanalytical system for micromorphological descriptions of the inorganic part of the soil material was developed in Australia by R. Brewer and J. Sleeman (1960) and later published by Brewer in his book *Fabric and Mineral Analysis of Soils* (1964a) (reprinted in 1976). This was the first attempt ever to establish a comprehensive system for making systematic and detailed micromorphological descriptions of soils. Although partly inspired by the morphoanalytical approach of Kubiëna, Brewer's system was based mainly on the experience of the author, who was interested in soil mineralogy. For this reason, the system was largely restricted to the mineral part of the soil. Barratt (1969) and Bal (1973) made extensions for the organic part.

Brewer's system was intended to be based on purely morphological criteria. However, one of its basic concepts, namely the plasma- skeleton grain concept, has a genetic base. Plasma and skeleton grains are not only defined by their absolute size (respectively smaller and larger than 2 μm), but also by their stability (See also Section 7.1 Background). This creates problems, as for example the case of minerals like calcite or gypsum, which can be stable in arid soils but will dissolve in the humid tropics. In his later publications (Brewer and Pawluk, 1975; Brewer and Sleeman 1988), the author almost abandoned these concepts. A most important contribution was the introduction of the concept of pedological features (Brewer and Sleeman, 1960, Brewer, 1964a), which by definition are those components that form by soil processes, such as clay coatings and Fe–Mn nodules. However, features inherited from the parent material, such as rock fragments or sedimentary structures, were also considered to be pedological features. Especially the fact that only single mineral grains could be part of the skeleton while compound grains (such as a quartzite fragment composed of two or more quartz grains) were considered pedological features, was felt by the users of the Brewer's system as problematic.

One of the merits of Brewer's system is that it made micromorphology more popular in many countries, especially in tropical and arid zones, where Kubiëna's system didn't provide concepts and terms for the description of fabrics. However, the greatest merit of the system is that it obliged micromorphologists to systematically analyze and describe all features of the soil thin section, as opposed to the morphogenetic system of Kubiëna which did not.

The second part of the 1960s showed an important expansion of soil micromorphology. Several new centers were created in Europe (e.g., in Great Britain, France, and Spain) and interest increased in the United States, Africa, South America, and Asia. The Post-Graduate Training Centers of Gent and Wageningen, and later also that of the ORSTOM (Paris), began attracting many students from Africa, Asia, and South-America and influenced this expansion. As a result, the knowledge on the micromorphology of soils increased sharply, forcing scientists to adapt the system, where possible, to new observations, adding new terms or changing or extending some of the concepts. Because this sometimes led to confusion, an international Working Group on Soil Micromorphology was created during the Third International Working Meeting on Soil Micromorphology, held in Wroclaw, Poland, in 1969. The purpose of the Group was to create an internationally acceptable terminology and classification. The result was the publication by Bullock et al. (1985) of the *Handbook for Soil Thin Section Description,* under the auspices of the International Soil Science Society (ISSS, now IUSS). The system of Bullock et al. (1985) became widely used by soil micromorphologists and it was later reworked for the first edition of this book (Stoops, 2003).

In 1984, FitzPatrick published his *Micromorphology of Soils.* It emphasizes the interpretation of soil thin sections, and not terminology. This is also the case for *Soil Microscopy and Micromorphology* by the same author (1993).

Micromorphology, as applied in the United States, is a tool rather than a discipline (Wilding and Flach, 1985; Wilding, 1997). On the contrary, in Europe (including Russia) and in Australia, micromorphology is often considered a discipline, and several research institutes and universities may have had one or more full-time micromorphologists on their staff. In several universities, micromorphology is still a regular part of the curriculum. These different approachs explain why scientists in the United States have contributed relatively less to the formulation of concepts and terms in the field of micromorphology, which is not to say that their work has been less important for the development of a description system. The efforts of a number of American soil scientists (including staff members of USDA and Soil Management Support Service of the USAID) in elaborating and refining U.S. system of soil taxonomy contributed to a better understanding of the distribution and genesis of micromorphological features, and as such to their interpretation and description. This effort is well illustrated in SSSA Special Publication 15, *Soil Micromorphology and Soil Classification* (Douglas and Thompson,

1985) and in Douglas (1990). The input of the U.S. soil micromorphology community is rather situated on a higher level of description, where no longer individual features are described, but global aspects are highlighted in function of soil processes, such as pedoplasmation (Buol and Weed, 1991) and redoximorphic features (Vepraskas, 1992).

In several centers of the former USSR (e.g., Dokuchaev Institute, Universities of Moscow and St. Petersburg), micromorphological investigations were performed. Most Soviet scientists followed the traditional terminologies of their petrographic school, although specific terms (e.g., polynite, clay pseudomorph) have been introduced (e.g., Parfenova and Yarilova, 1965). They also partly followed the system of Kubiëna, and even proposed some changes and additions. Later, concepts of Brewer were also introduced. A good review of Soviet scientists' concepts and terms is given in *A Methodological Manual of Soil Micromorphology* (Dobrovol'ski 1983). Since the 1990s the concepts of Bullock et al. (1985) and Stoops (2003) are commonly used in Russia.

Up to now, micromorphological classification systems and terminologies involve only the description of basic constituents of the soil (e.g., sand, clay) and combinations of these constituents. Such descriptions tend to be very lengthy and micromorphological features remain unrelated. A more efficient approach would be to describe, with one or few terms, a characteristic combination of features observed in a horizon or material, in the same way Kubiëna (1953) did for terrestrial and semiterrestrial humus horizons, such as tangel humus, dy, and gyttja. Also Kubiëna's concepts of Braunlehm, Braunerde, Rothlehm, and Roterde can be considered as higher levels of description. A first rather unsuccessful attempt to define higher levels of classification of soil fabrics was made by Brewer (1983), and developed by Brewer and Sleeman (1988), who combined several descriptive terms to produce long newly coined terms, without a clear relation to genetic horizons. Other proposals for higher levels of classification of soil fabrics were presented by Gerasimova (1994) and by Stoops (1994), but without success.

A review of the evolution of concepts and terminologies used in soil micromorphology was given by Stoops and Eswaran (1986) and Stoops (2009a). Miedema and Mermut (1990) prepared an annotated bibliography covering the period 1968 to 1986. Jongerius and Rutherford (1979) published a glossary of micromorphological terms.

2.3 STEPS OF MICROMORPHOLOGICAL ANALYSIS

A normal micromorphological study consists of different successive steps: sampling, preparation of thin sections, analysis and description of the thin sections, and finally the interpretation of the features observed.

2.3.1 Sampling

Sampling techniques are critical to any micromorphological study. The purpose of sampling is to obtain information relating to solving a particular problem, or to extrapolate information gained to understanding of other similar materials or soils. Thus errors made in sampling could influence subsequent interpretations.

Soil samples selected for micromorphological study should be representative of the area. A complete soil profile description should be prepared before sampling, so the location of the sample can be documented. Generally, the sample should be taken at the midpoint of the horizon and not across horizon boundaries, unless a horizon boundary is an item of investigation. Samples of the A1 horizon should include the uppermost part of the top horizon (e.g., the litter layer), unless the overlying material is sampled separately. Undocumented grab samples should not be taken except where they are taken to illustrate a particular feature or mineral assemblage. The same rules apply when sampling regoliths (e.g., saprolites) or archaeological strata (see also Stoops and Nicosia, 2017 and the references therein).

Soil samples for micromorphological study should be collected and packed in such a way that they are undisturbed and remain undisturbed while being transported to the laboratory (Murphy, 1986, Fox and Parent, 1993 and Fox et al., 1993). Since soils and regoliths are characterized by a vertical anisotropy, the vertical orientation of the samples should be clearly marked. When studying slope sediments, also the orientation in the landscape should be indicated in the sample. Knowing the orientation of the thin section may be an important aid to interpretation.

2.3.2 Preparation of Thin Sections

Sample preparation is as important as sampling. Details of preparation will not be covered here and the reader is referred to the excellent manual of Murphy (1986), and more recent publications by Fox and Parent (1993) and Fox et al. (1993). However, it is worthwhile to mention that the methods used to prepare thin sections should not introduce artifacts in the thin section. Some of these artifacts are addressed in Chapter 9.

2.3.3 Analysis and Description of Thin Sections

The study of thin sections comprises the use of different microscopic techniques and the recording of the data, which will be addressed in the following chapters. Despite the fact that both hardware and software became easily available to most scientists during the last decades, no serious progress has been made in storing and handling micromorphological data in a database.

2.3.4 Interpretation and Reporting

The interpretation of features observed and described in thin sections is beyond the scope of this book. The investigator must interpret the features observed, taking also in consideration the information available from field, laboratory analyses of the soil, and data from the literature. Useful manuals are, for instance, FitzPatrick (1984 and 1993), Stoops et al. (2010, 2018b) and Nicosia and Stoops (2017).

3. Aspects and Techniques of Thin Section Studies

3.1 FROM A TWO-DIMENSIONAL OBSERVATION TO A THREE-DIMENSIONAL REALITY

3.1.1 Introduction

The study of soil thin sections with a polarizing microscope presumes a working knowledge of optical mineralogy, some petrography, and frequently, some botany. Moreover, one needs a general understanding of the problems caused by observing three dimensional objects in two dimensions, and others related to the thickness of the thin section. This section provides the background information necessary for making accurate descriptions of features seen in thin sections.

3.1.2 Transition from Two to Three Dimensions

It is practically impossible to deduce a body's three-dimensional shape from one two-dimensional section through the body. One of the best illustrations of this is given in Fig. 3.1, where the shape of some possible sections through a cube (e.g., a pyrite crystal) are shown (e.g., a triangle, a square, a rectangle, a pentagon, a hexagon) and compared

Table 3.1. Three-dimensional objects and corresponding shapes as seen in two-dimensional sections.

Three-dimensional body	Two-dimensional section
equidimensional elongate platy lath like	equidimensional or prolate, according to plane of section
spherical	Circular
lenticular or ellipsoidal	Circular to oval

Guidelines for Analysis and Description of Soil and Regolith Thin Sections, Second Edition. Georges Stoops.
© 2021 Soil Science Society of America, Inc. Published 2021 by John Wiley & Sons, Inc.
doi:10.2136/guidelinesforanalysis2

with other geometric bodies. Other examples of common geometric bodies and their corresponding sections are listed in Table 3.1.

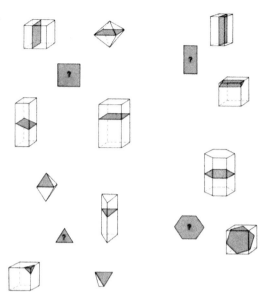

It is thus evident that only when a population of randomly-oriented, similarly-shaped bodies are present in a thin section, can one gather sufficient information to deduce the real shape of the object. One must also be aware that in many cases the observed section through a feature is a (sub)tangential one. This can sometimes give rise to erroneous interpretations. Sections through tubular channels, for example, mainly appear as circular

Fig. 3.1. Squares, rectangles, triangles, and hexagons can be created by sections trough a cube, or through other geometric bodies.

or oval pores, seldom as elongated ones (Fig. 3.2.). A tangential section through a nucleic nodule (i.e., a nodule formed around a core) may not hit the nucleus, so that a typic nodule seems to be present (Fig. 3.3.a). A subtangential section of a channel- or vugh hypo- or quasicoating (see Chapter 8) may look like an impregnative nodule (Fig. 3.3.b); a tangential section through a bended channel coating may appear as an infilling (Fig. 3.3.c); a cross-section through a channel infilling with crescent fabric will appear as a concentric nodule. Deep embayments in minerals (e.g., in quartz) (sub)perpendicular to the plane of thin section may appear as holes in the grain (Donaldson and Henderson, 1988) (Plate 6.3e and f).

It is statistically impossible for all grains in a thin section to be cut through their largest diameter (Fig. 3.4.). Their size in thin section is therefore in general somewhat smaller than their actual size. From a statistical

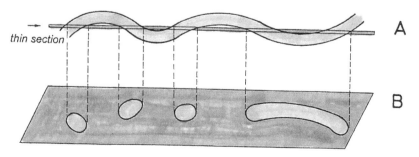

Fig. 3.2. Possible sections through a tubular channel (after Stoops, 1978a). (A) Thin section seen from aside, (B) thin section seen from above.

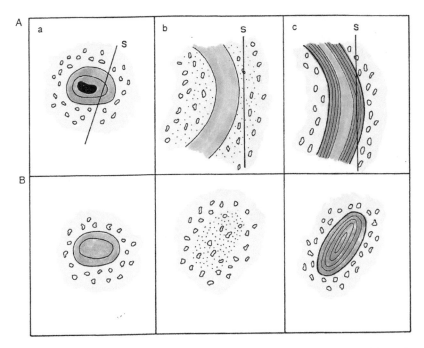

Fig. 3.3. Effects of thin section location on feature appearance. (A) Different sections through the center of: (a) a nucleic nodule, (b) a hypo-coating, and (c) a clay coating. (B) Tangential section according to S.

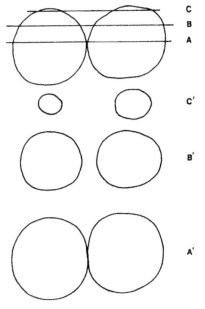

Fig. 3.4. Sections through spherical grains. Lines A, B, and C are sections cut through the spherical particles at different points. The resulting sizes and shapes of particles in the section are shown in A', B' and C'. Only the section through the center (Line A) shows the real diameter in a thin section (A'). The likelihood of cutting two or more grains at their point of contact is small: therefore even touching grains generally appear to be floating (section B' and C').

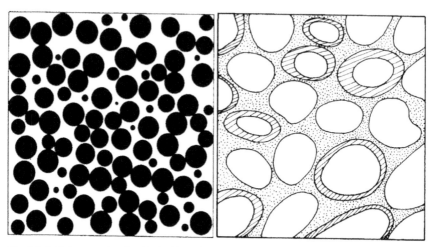

Fig. 3.5. (left) Apparent size of grains in thin section. A packing of spheres with actual size of 1 mm shows an apparent mean size of 0.785 mm (from Harrell, 1984).

Fig. 3.6. (right) Packing of grains in clay mass seen in polarized light (PPL) and between crossed polarizers (XPL). The white grains correspond to the image seen in PPL; the striped rim is the part that becomes visible in XPL.

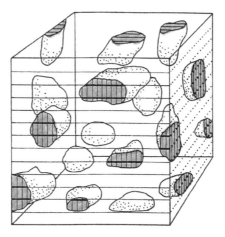

Fig. 3.7. Relation between volumes, areas and lengths and frequency of points. Spherical bodies (A) and the matrix (B) are shown. The volume ratio A/B is proportional to the area ratio A/B in a section, which is proportional to the length ratio of the sum of the line segments through A and B (front surface) or to the ratio of the number of points covering A and B (right surface).

standpoint, the average grain diameter of larger grains (significantly larger than the thickness of the thin section) measured in thin sections is 0.785 times the actual diameter. As a result, a random plane section through a set of grains having an actual diameter of 1 mm will show an apparent mean size of 0.785 mm (Adams, 1977; Harrell, 1984) (Fig. 3.5). Harrell (1984) and Longiaru (1987) published charts for the visual comparison of different degrees of sorting. Statistical methods have been proposed to determine the relationship between grain sizes measured in thin sections and those determined by sieving (Harrell and Erikson, 1979, see review by Johnson, 1994). As the particles decrease in size, the measured value is more like the actual value. For example, an infinite number of particles having a diameter of 20 μm in a section of 30 μm thick would be measured at 19 μm, while grains of 200 μm would be measured at 167 μm.

For the same reason one seldom sees coarse grains in contact with each other in thin sections, as the points of contact are mostly outside the section plane (Fig. 3.4.B, C). A close packing of coarse sand grains therefore appears in thin sections as a swarm of isolated grains, occurring at some distance from one another (Plate 4.4a and b; Plate 4.5a). However, the effect is counterbalanced by the thickness of the thin section, especially when observations are made in XPL; not the image of one plane of observation is seen, but the integration over the whole thickness of the section (Fig. 3.6), resulting in seemingly larger sizes (see also Section 3.1.3 and 3.1.5) (Harrell, 1981). This is also the case when using modern multifocal digital microscopes.

The problem of observed sizes in thin sections versus real sizes in three dimensions occurs of course also when measuring other features than grains, such as pores. Its importance in the evaluation of porosity in soil thin sections was already discussed by Dorronsoro et al. (1978b).

The transition from a three-dimensional feature to a two-dimensional appearance will also influence the apparent connectivity. One continuous irregular unit may appear as several discontinuous units in thin sections, a channel for example, as shown in Fig. 3.2.

The ratio between the volume of different constituents, when arranged at random in a three-dimensional body corresponds to the ratio of their areas in a random section (law of Delesse) (Weibel, 1979).

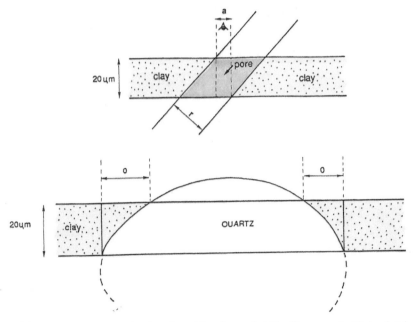

Fig. 3.8. Wedging effects: **(A)** section through an oblique fissure of width *r*. The apparent width is reduced to a *a'* when the section is viewed under a microscope. **(B)** Tangential section through a quartz grain showing a colored fringe (o) where it is overlapped by clay.

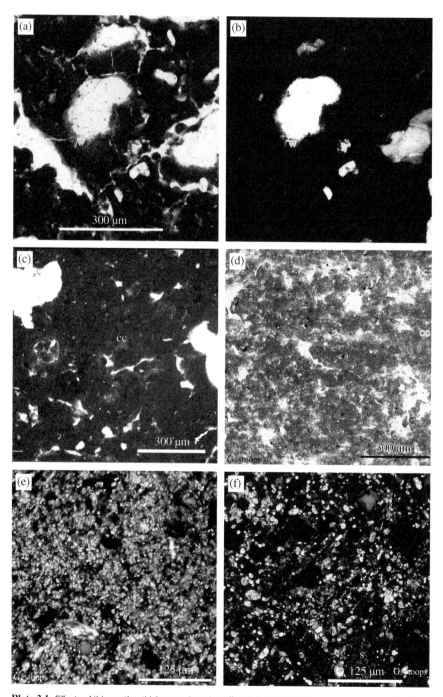

Plate 3.1. Effects of thin section thickness. **a)** wedge effect (w) on border of quartz grain embedded in speckled dark brownish clay, partially covered by the clay (PPL); **b)** same (XPL) **c)** reddish micromass and fragments of limpid clay coatings [lighter, limpid spots (cc)] in tropical soil; thin section at normal thickness. (PPL); **d)** same thin section, but thinner: note change in color and continuity of micromass caused by thinning of section (PPL); **e)** calcite crystallites with high interference colors overlap in calcitic crystallitic b-fabric at normal section thickness (XPL); **f)** same thin section but less thick: the area occupied by calcite crystallites is strongly reduced, and zones with a speckled b-fabric appear (XPL).

Moreover the ratio of these areas corresponds to the ratio of the sum of their segments along a line (Fig. 3.7.). These mathematical relationships are used in stereology to calculate relative volumes of rock, soil, or regolith constituents (e.g., porosity) on the basis of microscopic measurements of thin sections.

3.1.3 Wedging Effects

A thin section has a thickness ranging mostly between 20 and 30 μm. As most boundaries between constituents will be oblique to the plane of the section, a wedge-shaped thinning, resulting in a partial overlapping of two constituents near their boundaries is to be expected. This effect is mostly visible if one of the constituents is transparent and sufficiently large, such as a pore or a quartz grain. As a result, pores, especially fissures, may seem to be much smaller than they really are (Fig. 3.8.A). In the same way, transparent grains also seem smaller. This results in an overestimation of clay in thin sections, as shown by Murphy and Kemp (1984). On larger grains (e.g., quartz or calcite grains) embedded in a clay matrix, a dusty or colored fringe appears (Fig. 3.8. B, Plate 3.1a and b). The latter feature has sometimes been misinterpreted as a proof for secondary growth, when the outline of the grain is clear, or it can suggest a granostriated b-fabric (see Section 7.2.4.2). Due to the wedging effect, mineral grains can show lower interference colors along their borders, a feature that has been mistaken frequently for peripheral weathering).

A progressive thinning is also frequently observed at the border of the thin section, although this should not occur in well-made thin sections. In the clay matrix, this is generally visible by a change in color (from dark to light or from reddish to yellowish), an increase in limpidity, and an increase of areas showing interference colors resulting from clay orientation (Plate 3.1c and d). If the matrix consists of clay with a high content of microcrystalline calcite, the calcite particles are overlapping at standard thickness, whereas they become more and more separate with decreasing thickness of the section (Plate 3.1e and f).

3.1.4 Minimal Visible Size and Magnification

In optical microscopy, the minimal observable size in principle depends on the magnification used and the quality of the optical system. In the case of thin section studies, the thickness of the section also plays a role, as well as the contrast between the particle and its surroundings.

In practice, the minimal size for strongly contrasting isolated particles (e.g., opaque grains in a clay matrix) to be visible with objective 50 (the highest magnification normally used in micromorphology) is about 3 μm. Individual, noncontrasting particles with diameter < 20 μm will be difficult to distinguish from one another (e.g., in fine silt) because of overlapping. This means also that smaller features are not detected; hence it is important to distinguish clearly between "not detected" and "absent" in descriptions.

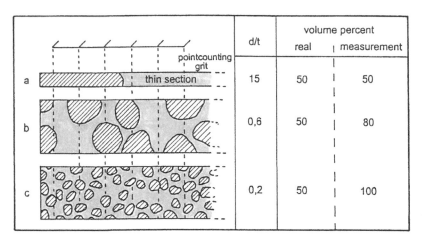

	d/t	volume percent	
		real	measurement
a	15	50	50
b	0,6	50	80
c	0,2	50	100

Fig. 3.9. Influence of the Holmes effect on area determination in transmitted light (after Maier-Kühne and Babel, 1984). d/t is the ratio between grain diameter (micrometer) and thickness of the section (micrometer). The real volume percentage is determined by point counting in oblique incident light (OIL); measured volume percentage obtained by point counting in transmitted light.

3.1.5 Holmes Effect

Taking into account the effects mentioned above, it is clear that a size difference will exist between features observed in transmitted and in reflected light, as the latter shows the size in one plane and the former the integrated size covering the thickness of the section. Opaque grains, for instance, will always appear larger in transmitted light than in reflected light, whereas for fine pores the opposite is true (Holmes, 1927). This phenomenon is called the Holmes effect. It plays an important role in all stereometric determinations in thin sections, as it may give rise to a considerable overestimation of dark or opaque particles, smaller than the thickness of the thin section (Maier-Kühne and Babel, 1984) (Fig. 3.9).

3.1.6 Orientation

Interpretation of the orientation of elongated features (e.g., planar voids, prismatic minerals) performed on two-dimensional images is difficult because the angle between the plane of the section and that of the object is usually not known. For instance, a fissure with an inclination of 45° will appear as 45° in a vertical section perpendicular to the dip direction, but will have little inclination in a section parallel to it (compare the dip of a layer in a quarry). For randomly oriented sections, all possible values between 0° and the actual inclination can therefore be found.

3.1.7 Optical Illusion

Scientists should be aware also that our eyes and brains are subject to perception: what we see in not always the reality. A well-known example is that of two perfectly parallel lines that, because of the context, appear no longer parallel. This concept of perception (optical illusion) has been used to explain differences in results of optical observations and image analysis, especially in the case of b-fabrics (Zaniewski, 2001, Zaniewski and van der Meer, 2005).

3.2 MICROSCOPIC TECHNIQUES FOR THIN SECTION STUDIES

3.2.1 Introduction

Most micromorphological analyses of thin sections are performed with the aid of a petrographic microscope. Observations are done alternatively in polarized light (PPL) and between crossed polarizers (XPL). As this technique is considered to be standard, no further discussion is given here. A short introduction to some essential concepts, and references to manuals are given in Appendix 1.

However, this method is in some cases not sufficiently powerful to allow the identification, or even observation, of some specific components. Special techniques might be helpful in these cases. They include optical techniques (e.g., circular polarized light, UV-fluorescence or cathodoluminescence), extraction and bleaching techniques, and staining techniques. Kubiëna (1938) already recommended staining of some soil components. For details of the physical and/or chemical basis of the techniques and equipment explained, the reader is referred to works cited. An excellent general handbook on this topic is Hutchison (1974). The reader is also referred to Drees and Ransom (1994) for the description of microscopic techniques used in soil science.

Microanalytical techniques, such as WDXRA, EDXRA, LAMMA, micro XRD, FTIR-microscopy, and microdrilling on uncovered thin sections have over the past few decades partly replaced some of these optical techniques; their discussion is beyond the scope of this book. Several introductions to the use of these techniques in soil micromorphology are presented in Nicosia and Stoops (2017). Methods such as microprobe analysis that yield total chemical data cannot always replace such techniques as selective extraction of different Fe components in thin sections.

As a general rule it is strongly recommended that thin sections should be studied and described in detail, and if possible photographed, before using other techniques. This is especially true for techniques altering the material in the thin section (e.g., selective extraction, staining). In many cases conclusions can only be reached by comparing the spot before and after the treatment, or by comparing the results of different observation methods.

3.2.2 Optical Techniques

3.2.2.1 Introduction

The optical methods described here are limited to a few techniques, which do not require sophisticated equipment, difficult preparation techniques, or a specialized knowledge of optics. Methods such as phase contrast (Altemüller 1964, 1997), interference microscopy, and confocal microscopy are not discussed here.

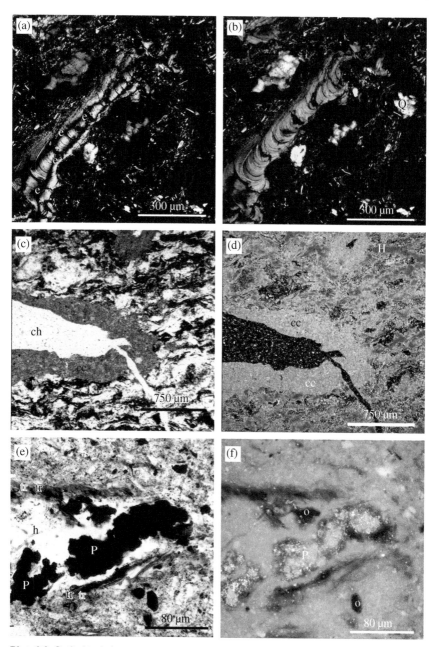

Plate 3.2. Optical techniques. Circular polarized light (CPL): a) dense complete infilling of channel with fine clay with continuous orientation, as demonstrated by the two sharp parallel extinction lines (e) and strong interference colors (XPL); b) same, but in CPL; notice absence of extinction lines in the infilling, and its clear crescent fabric; the quartz grain (Q) at right that was in extinction in XPL, is now clearly visible (CPL). Transmitted dark field illumination: c) uniform typic coating (cc) of speckled yellowish coarse clay on channel (ch) in stratified saprolite (PPL); d) same, the black streaks observed in PPL become reddish (fine dispersed hematite), the white kaolinite dull black and the yellow clay coating (cc) waxy yellow (fine goethite admixture) (TDFI). Oblique incident light: e) channel with tissue residues (part of a root), colorless hyphae (h) and aggregates of pyrite framboids (P) as loose discontinuous infilling in Acid sulfate soil (PPL); f) same,the strongly humified organic particles (o) are dull black; the pyrite (P) has a yellowish metallic luster (OIL).

Using a normal polarizing microscope, a few simple tricks can be used to obtain information in an easier way, or to get extra information. Most microscopes allow observation between partially crossed polarizers, by rotating either the polarizer or the analyzer a few degrees. Voids will appear as gray areas. An example of its application is shown in Plate 4.5b; in PPL coarse grains would be white and difficult to distinguish from the voids; in XPL grains would show interference colors, but the aggregates would be black even as the voids. Partly crossed polarizers allow observing the three components simultaneously. In most cases a similar effect can be obtained by inserting the 1/4 λ retardation plate (mica compensator).

3.2.2.2 Circular Polarized Light

For normal observations between crossed polarizers so-called linear polarized light is used. Under those conditions, minerals show extinction every 90°, when their vibration direction is parallel to that of the polarizer or analyzer. To overcome this, so-called circular polarized light is used.

The equipment for this technique consists of two identical 1/4 λ retardation plates (also called mica plates), one in the compensator slot (i.e., between the sample and the analyzer), and the other in the filter slot of the condenser (i.e., between the sample and the polarizer) making an angle of 45° and 135° respectively with the analyzer and the polarizer. They are mounted in such a way that the slower ray of the upper plate matches the faster ray of the lower plate, and vice versa. Phase differences are completely annihilated with this arrangement of plates. For observation, the analyzer should be inserted. As a result anisotropic grains will only be in extinction when the section is perpendicular to their optical axis (Plate 3.2a and b). Isotropic and opaque bodies, as well as voids, will remain black. Because some important information will be lost with this observation technique (e.g., extinction orientation of minerals, extinction lines in clay coatings), sections should always also be studied in PPL and XPL.

This technique is particularly useful when extinction phenomena of anisotropic grains have to be avoided, such as for porosity studies (Ruark et al., 1982), and to study clay orientation (Jim, 1988b) and b-fabrics (Pape, 1974) (Plate 3.2a and b; Plate 7.3c and f), for instance in the case of their quantification.

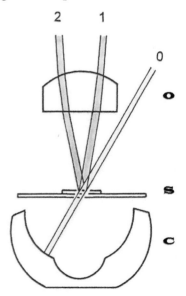

Fig. 3.10. Optical path in darkfield condenser. The direct beam 0 passes the objective. Only when it is deflected by a sample (1 and 2) does the light enters the optical system of the microscope. The condenser (C), sample (S), and the objective (O) are shown.

3.2.2.3 Dark-field Illumination

For normal observations on thin sections, brightfield microscopy is applied, using orthoscopic or conoscopic light with a full beam. A darkfield condenser (mirror

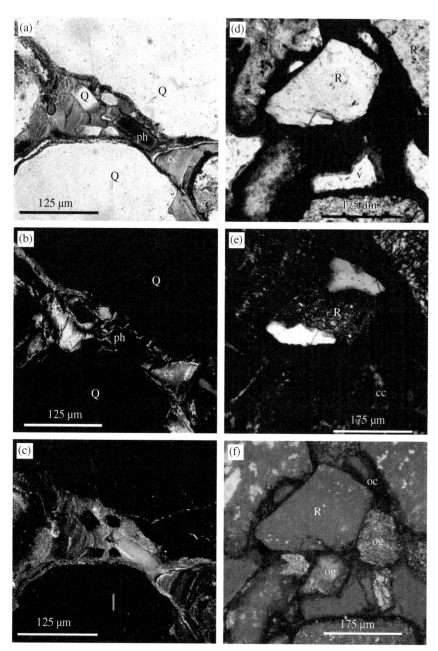

Plate 3.3. *Oblique incident light.* **a)** bridge between two rounded quartz grains (Q) consisting of limpid and speckled strongly oriented clay (cc), some fine quartz grains and (in the center) a purple brown rectangular phytolith (ph) (PPL); **b)** same, note interference colors due to strong orientation of the clay (cc) and the isotropism of phytolith (ph) (XPL); **c)** same, note black color of quartz, waxy appearance of clay and white opaline shine of phytolith (TDFI); **d)** rock fragments (R) in opaque micromass and brown typic clay coating on void (v) (PPL); **e)** same, the rock fragments (R) can be identified as chert; the abnormal interference colors of the coating (cc) show that it consists of strongly oriented fine clay and iron hydroxides (XPL); **f)** same, part of the opaque mass consists in fact of opaque grains (og) with metallic luster (probably magnetite or ilmenite), whereas the rest (mainly present as coatings) (oc) consists essentially of dull Mn (hydr)oxides; (OIL). The micrographs of these two sequences show a specific aspect: the "observed fabric"; their combination yields an "integrated fabric", necessary for interpretation.

condenser) consists of a concave lens with an opaque circular field in the center. As a result, a shallow, hollow light cone is formed. If no object is in its path, the light bypasses the objective, and the field of view is dark (Fig. 3.10). An object will deflect the light, and part of this will reach the objective. Transparent particles with low contrast will have their contour illuminated in this technique.

In micromorphology, the method is useful for the study of the fine mass (either in the groundmass or in pedofeatures) (Plate 3.2c and d; Plate 8.9b). The obtained image is similar to that in oblique incident light, except that the mineral's luster, an important characteristic of opaque minerals, cannot be determined.

3.2.2.4 Oblique Incident Light and Dark Ground Incident Light

Opaque particles (e.g., pyrite, charcoal) or features (e.g., Mn-nodules) cannot be studied in transmitted light. The appropriate method is to use reflected light (also called ore microscopy), as done by geologists. A light beam passes through the objective and after reflection on the mineral, it passes again through the same lens (Fig. 3.11a). The identification of minerals is based on color, luster (measured quantitatively), polarization effects, and hardness (quantified by measuring the diameter of a cone produced by a diamond pointed weight) (Galopin and Henry, 1972, Jambor and Vaughan, 1990, Peckett, 1992). The drawback of this method is that apart from a special microscope and accessories, including a photometer, only highly polished, uncovered sections can be used. These requirements make the method unsuitable for routine studies of soil and regolith thin sections. Its application in micromorphology of archaeological materials is discussed by Ligouis (2017).

An alternative method, more suitable for identification of dark components in normal, covered soil thin sections, is using oblique incident light and dark-field incident light obtained by a vertical dark-field illuminator (e.g., Ultropak, E. Leitz, Wetzlar, Germany). In the latter case the light passes through a coaxial condenser mounted around the objective and forms a cone whose apex is located in the plane of the thin section, right beneath the objective (Fig. 3.11b). The method does not allow an exact mineralogical determination of different ore minerals, but is sufficiently

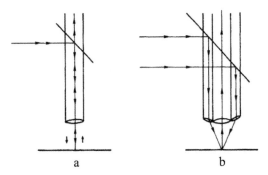

a b

Fig. 3.11. Optical path for (a) reflected (direct incident) light (ore microscopy) and (b) oblique incident light.

Table 3.2. Characteristics of some of the most important minerals in oblique incident light †

Component	Transmitted light	Oblique incident light	
		color	luster
Charcoal	opaque	black	dull
Mn oxides	opaque	black	dull
Hematite			
massive	opaque	dark gray	metallic
fine	red	red	dull
Goethite			
massive	reddish brown	black	metallic
fine with clay	brown-yellow	yellowish green	dull
Pyrite	opaque	green-yellow	metallic
Chalcopyrite	opaque	yellow	metallic
Magnetite	opaque	gray	metallic
Quartz	colorless	dark	
Clay	gray	(white)	waxy
Jarosite	yellow	bright green	waxy
Opal	gray	blue-green	bright

† Data from Van Dam and Pons (1972), and Stoops, unpublished data.

precise for most minerals occurring in soil thin sections. When these special accessories are not available, similar results can be obtained with a light source mounted above the microscope, especially a cold light with a single- or double-arm swan neck glass fiber light guides. An inexpensive solution is a simple, led torch, if possible with variable focus, which facilitates concentrating the light on a relative small area of the section. Care should be taken to switch off the transmitted light, or to cover the built-in illumination with an opaque cover. It is recommended to lower the condenser as much as possible, to avoid reflectance of the incident light on it. Plate 3.2e and f; Plate 3.3 and Plate 6.11a and b illustrate the use of oblique incident light to distinguish different opaque constituents.

Table 3.2 gives some helpful information for the identification of soil constituents with oblique incident light (OIL). Whereas many identification tables are published for reflected light (ore microscopy), no systematic key yet exists for OIL. Dark ground incident light equipment is also useful for fluorescence microscopy (see Section 3.2.2.5.).

3.2.2.5 Fluorescence Microscopy

3.2.2.5.1 Introduction

Certain substances, when irradiated with short-wave radiation (UV, violet, blue) emit radiation with a longer wavelength. This phenomenon is called luminescence. Two types are distinguished. When the emission persists for some time after the irradiation has stopped, it is called phosphorescence; when it persists only as long as excitation continues, it is called fluorescence.

Two types of fluorescence are important in earth science:

Primary fluorescence or autofluorescence. The specimen exhibits the phenomenon of fluorescence without pre-treatment. Well-known examples in mineralogy are fluorite, calcite, gypsum, and most U minerals such as autunite. In soil micromorphology, the primary fluorescence of cellulose is a most useful diagnostic for fresh plant residue, even as the autofluorescence of bones and phosphates. Primary fluorescence has a high diagnostic value in coal petrography.

Secondary fluorescence. Certain materials can be stained with fluorescent dyes and will then exhibit secondary fluorescence when irradiated. Such dyes are called fluochromes. Examples include acridine orange, auramine, coriphosphin and rhodamine. Some of them are used for selective staining of some types of biological materials. Clays can also be stained by fluochromes (see also Section 3.2.4.5).

3.2.2.5.2 Equipment

Specially-designed fluorescence microscopes are commercially available, but in the case of petrography or micromorphology, the adaptation of a research polarizing microscope is recommended. The simplest equipment consists of an exciting filter that blocks all but the exciting wavelengths, and a suppression filter to stop the exciting radiation to produce a dark background (in the case of UV-light it also protects the eye). The exciting filter is mounted between the light source and the specimen (generally in the light housing), while the suppression filter is placed in a special slot between the specimen and the ocular. In earth science research, the broadband excitation system is used. This means that the exciting light is filtered out with the aid of dyed glass filters. For an optimum yield of fluorescence, an exact combination of filters is important. Table 3.3 gives the examples of combinations of filters.

In soil micromorphology oblique incident light (see Section 3.2.2.4) is normally used for fluorescence studies (Altemüller and Van Vliet-Lanoë, 1990; Stoops, 2017). A special vertical illuminator (e.g., Ploemopak, Leitz Wetzlar, Germany), contains the appropriate filters and dichroic beam splitters. The advantage of incident light is that no losses occur through scattering or primary absorption in the specimen or in the support glass, and a dark background can be obtained.

A combination of simultaneous incident- and transmitted-light fluorescence (using different wavelengths) can be useful to distinguish different types of fluorescence (e.g., different stainings, primary and secondary fluorescence) (Hartmann et al., 1992). The light source must yield sufficient radiation in the short range. A normal tungsten lamp is

Table 3.3. Combination of exciting and suppression filters.

Exciting radiation	Exciting filter	Suppression filter
		nm
Ultra violet	330–380	420
Blue violet	400–440	480
Blue	420–490	520

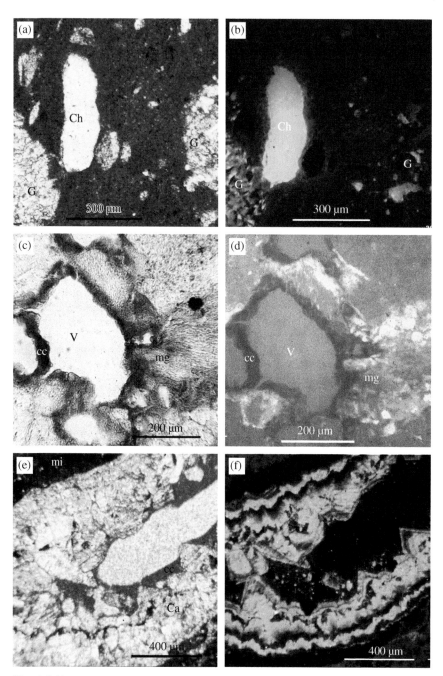

Plate 3.4. *Fluorescence microscopy*. **a)** empty channel (Ch) and loose continuous channel infillings of fine lenticular gypsum (G), in gypsiferous soil (PPL); **b)** same, note blue fluorescence of resin in channel (Ch) and in packing voids of gypsum infilling (UVF); **c)** coating (cc) of Fe-stained colloidal material around void (v) between weathered mineral grains (mg) in Bir of Spodosol (PPL); **d)** same, Fe-free aluminum hydroxide colloids show yellowish fluorescence, whereas colloids with Fe and/or organic matter (cc) do not show fluorescence (BLF); **e)** sparitic calcite coating (Ca) in micritic (mi) calcite nodule, with juxtaposed yellowish clay coating (cc) (PPL); **f)** same; the different steps of calcite crystallization of the coating are clearly expressed by differences in cathodoluminescence intensity (CL).

generally not suitable and a halogen lamp is only slightly better. A 50 W high-pressure mercury lamp is sufficient for normal work, but a more intense light (e.g., 50 to 200 W mercury [HBO] or 75 to 100 W Xenon high-pressure lamp [XBO]) is necessary for the study of objects with low fluorescence intensities (Stoops, 2017). Care should be taken to work in a relatively dark room, especially when fluorescence is weak, and to avoid lateral light on the thin section.

Autofluorescent resin should be avoided for mounting, because it may produce an inconvenient background. Poorly polymerized resin is often fluorescent (e.g., at the border of the thin section or around air bubbles, due to the availability of oxygen).

Other errors that can occur in fluorescence microscopy are: (i) apparent nonfluorescence caused by insufficient excitation or transmission, (ii) extraneous fluorescence caused by the preparation technique, and (iii) internal reflection or refraction of fluorescent light (Cercone and Pedone, 1987).

3.2.2.5.3 Application

Porosity. Pores can be made visible by adding a fluochrome to the resin used for impregnation of the rocks, sediments or soils (Werner, 1962, Dorronsoro et al., 1978a, Yanguas and Dravis, 1985). In general, this fluorescence is much stronger than the autofluorescence of the soil constituents, and therefore the interference of the latter is limited (Plate 3.4a and b). Hartmann et al. (1992) recommended the simultaneous use of transmitted and incident UV light for the description of microstructures, using fluorescent resins.

Minerals. Although fluorescence cannot be used as a method for mineral determination, it may be very helpful in fabric studies. The minerals that always show fluorescence (such as many U minerals) are never main constituents of soils. Several other, more common, detrital or pedogenic minerals will be fluorescent if they contain the correct activators because the minerals themselves do not fluorescence. If they are pure, they are nonfluorescent.

The most common fluorescent minerals are some phosphates, especially apatite, and some carbonates. Bones give a greenish yellow fluorescence when excited with blue light, and a light blue fluorescence in UV-light (Plate 6.10). This is also the case for other materials, such as coprolites of carnivores and guano (Karkanas, 2017; Villagran et al., 2017; Karkanas and Goldberg, 2018).

The original fabrics of carbonate nodules or lithic fragments, obliterated by later recrystallizations, may become visible by fluorescence microscopy. Calcitic phytholiths are also often fluorescent (Altemüller and Van Vliet-Lanoë, 1990). Soil carbonates generally give a bluish color in UV–OIL and a yellow color in blue OIL (Dravis and Yurewicz, 1985) Different successive phases of formation of calcite coatings or nodules can often be recognized by their fluorescence.

Van Vliet-Lanoë (1980) has shown that a relatively strong fluorescence in some soils (e.g., saprolites, spodic horizons) is caused by free Al, present in a stage intermediate between Al ions (e.g., adsorbed on the clay) and large hydroxipolymers. Zones where alteration or Al accumulation occur can be recognized (Plate 3.4c and d; Plate 6.13c). The fluorescence is strongly reduced by the presence of Fe and/or polycondensed organic matter. Both UV and blue exciting filters were used, sometimes combined with a BG3 filter (250–450 nm) to decrease the autofluorescence of the resin. A greenish to greenish blue fluorescence is obtained with UV excitation.

Organic Matter. Babel (1972) discussed the use of UV fluorescence for the study of organic matter in soil thin sections. He recommended a 2-mm UG1 exciting filter in combination with a 470 nm suppression filter. Although transmitted light is in general not recommended because of a weak background fluorescence which decreases the contrast, he advised this method just because this background makes it possible to also observe other non-fluorescent features, which cannot be observed when oblique incident light is used. Due to the yellow color of the suppression filter, blue fluorescence colors become greenish. The yellowish brown color of humus components may, by absorption, also cause this color shift and considerably reduce its intensity. This problem can be overcome by bleaching and extracting humic components with a sodium hypochlorite treatment (see Section 3.2.3.4.). Fluorescent fabric elements, mainly cell walls, are not destroyed by this treatment.

The fluorescent fraction of organic matter mainly consists by cell walls. Fungal hyphae can also be strongly fluorescent. Newly formed colloidal organic matter (e.g., as found in grain coatings in spodic horizons) is not fluorescent, and can even suppress autofluorescence of other colloids (Van Vliet-Lanoë, 1980). Common materials that can be observed include:

Lignified tissues have a strong blue color when excited with UV. With increasing humification the color gradually shifts to whitish and yellow, and sometimes disappears.

Cutinized and suberized cells show a whitish fluorescence with UV; fungal hyphae behave the same way.

Parenchymatic tissues, when excited with blue light, have a brownish to reddish color.

Young cells in root tips and root hairs show a blue color with UV.

In addition to autofluorescence, secondary fluorescence has also been used for recognition and identification of organic matter in thin sections. Plant tissues can be stained with fluochromes (e.g., acridine orange). Altemüller and Vorbach (1987) developed following procedure (Altemüller and Van Vliet-Lanoë 1990):

- Clean uncovered thin section with benzene (care should be taken when using this cancerogeneous product).
- Stain with acridine orange, 1/10,000 in 10% (v/v) HCl, 2 min.
- 1 min. 10% (v/v)HCl.
- 1 min. distilled water, repeated.
- Dry with compressed air.
- Mount

Under blue excitation, the cell walls of roots and other plant tissues exhibit a green or yellowish green color, whereas the clay will appear orange to yellowish.

Instead of acridine, trypaflavine (= acriflavine) (1/10,000 in a 4% (v/v) formalin solution) can also be used, giving a strongly green fluorescence to root tissues.

Bacteria can be stained in thin sections using fluorescent brightener calcofluor white M2R (Postma and Altemüller, 1990).

Experiments on staining of organic matter in soils, prior to impregnation were done by Postma and Altemüller, 1990 (see also Altemüller and Van Vliet-Lanoë, 1990) and Tippkötter (1990), but are beyond the scope of this book.

3.2.2.6 Cathodoluminescence

3.2.2.6.1 Introduction

When a high-energy electron beam hits a solid surface, it causes several physical processes to take place, including the emission of secondary and back-scattered electrons, X-rays and visible light. The latter phenomenon is called cathodoluminescence (CL). The higher the energy of the beam, the deeper it will penetrate into the target. For beams in the 10 to 20 kV range a penetration depth of 1–2 mm is to be considered.

The color and intensity of CL are determined by the structure of the crystal lattice, which includes exsolved phases, defects, and impurities. The latter represents the most important factor, because the impurities act as activators. Activators mostly occur as substitutions in the crystal lattice, but sometimes they may be present as interstitial impurities. Important activators are Mn, Pb, and rare earth elements. Cathodoluminescence intensity is a function of the activator's concentration, but decreases after an optimum value is reached. In most cases, their optimal concentration is < 1000 ppm. Some elements perform an opposite role and reduce or inhibit the luminescence. The most important of these quenchers is Fe^{2+} (Marchall, 1988; Pagel et al., 2000).

3.2.2.6.2 Equipment

Although most scanning electron microscope (SEM's) are equipped with a device allowing the observation of CL (usually in black and white), special equipment that can be mounted on the stage of a polarizing microscope is preferred for thin section studies, because this allows combined PPL, XPL, and CL observations.

The CL microscope attachment basically consists of a metal vacuum chamber with two viewing ports of leaded glass (one on top for observation, one at the bottom for transmitted light). A built-in electron gun is oriented on the sample and provides a beam of a few hundred micrometers. Two types of electron guns are available: a cold cathode electron gun and a hot cathode electron gun. The latter is more powerful and used for the CL study of silicates, whereas the former is used mainly for carbonates.

3.2.2.6.3 Sample Preparation

Uncovered thin sections are usually perfectly suitable, but a highly polished surface will enhance CL intensity. The surface should be clean, because some products used for polishing may be luminescent. Heat developed as a result of the electron impact may melt some of the thermosetting resins used for mounting, but epoxy resins and polystyrenes will resist melting, although they may show some burning effects after long exposures to the beam (e.g., in the case of microphotography, where long exposure times are sometimes needed). Some hydrated minerals, such as gaylussite, are immediately decomposed by the beam.

3.2.2.6.4 Application

At this time, the two most common applications of CL in soil micromorphology are: (i) distinguishing varieties of the same mineral, for instance calcite of different origins (e.g., Khormali et al., 2006), and (ii) the observation of fabrics in carbonatic pedofeatures, for instance the presence of veins in calcite nodules, invisible with other optical methods as a result of recrystallization. Also, it can be used to distinguish detrital and newly formed authigenic components. Some examples are given:

Calcite: about 100 to 1000 ppm Mn^{2+} present as an activator gives a yellow, orange to red CL-color. Fe, Co and Ni act as quenchers. Rare earth elements are considered as activators. A red color is more typical for dolomite. Good results can be obtained with the cold cathode electron gun (Plate 3.4e and f).

Aragonite: yellow- green.

Sulfates: barite (activated by Pb) and celestite shows CL, gypsum does not.

Apatite shows different colors, mainly yellow (activated by Mn).

Silicates mainly show different grades of blue, but in most cases a hot cathode electron gun is required. Quartz is a special case: slightly different colors occur, related to the host rock's origin: blue violet for igneous rocks, brown for metamorphic rocks, greenish brown for hydrothermal deposits. Secondary authigenic quartz rims are not luminescent (Marchall, 1988).

3.2.3 Selective Extractions

3.2.3.1 Introduction

Selective extraction of components from thin sections has three objectives: (i) to identify constituents, (ii) to separate similar constituents (e.g., as in differentiating "amorphous" from "crystalline Fe oxyhydrates), and (iii) to make those features that are being masked by other mineral components visible (e.g., extraction of carbonates or Fe oxyhydrates from the groundmass to enhance the b-fabric). Extraction may play an important role as pre-treatment for some staining tests.

All extractions are done on uncovered thin sections. Careful washing with distilled water after each treatment is a must. If the binding to the glass is not good, there is a risk of the section peeling off, especially when

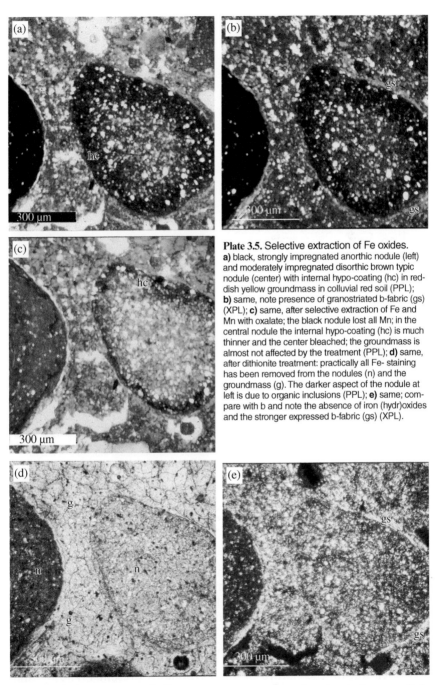

Plate 3.5. Selective extraction of Fe oxides.
a) black, strongly impregnated anorthic nodule (left) and moderately impregnated disorthic brown typic nodule (center) with internal hypo-coating (hc) in reddish yellow groundmass in colluvial red soil (PPL); **b)** same, note presence of granostriated b-fabric (gs) (XPL); **c)** same, after selective extraction of Fe and Mn with oxalate; the black nodule lost all Mn; in the central nodule the internal hypo-coating (hc) is much thinner and the center bleached; the groundmass is almost not affected by the treatment (PPL); **d)** same, after dithionite treatment: practically all Fe- staining has been removed from the nodules (n) and the groundmass (g). The darker aspect of the nodule at left is due to organic inclusions (PPL); **e)** same; compare with b and note the absence of iron (hydr)oxides and the stronger expressed b-fabric (gs) (XPL).

warm solutions are used. To compare treated and untreated areas, it is useful to cover part of the section with silicone grease. Color micrographs of the same spot before and after treatment are recommended for comparison. It is useful to apply a thin layer of glycerin on the uncovered thin section when making micrographs, to enhance the contrast.

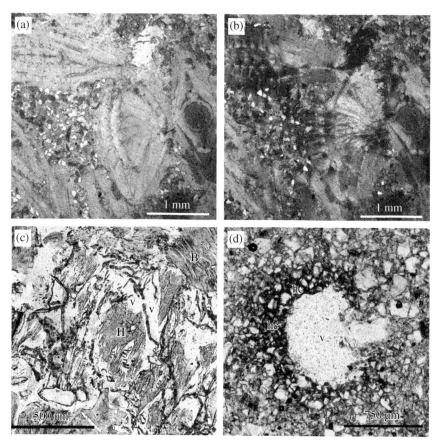

Plate 3.6. *Differential staining of carbonates.* **a**) nummulitic limestone stained with alizarine and fer-ricyanide; pink zones correspond to pure calcite, bluish zones to Fe-bearing calcite (PPL); **b**) same (XPL). *Staining of clay.* **c**) irregular banded alteration of hornblende (H) forming a boxwork of colorless smectitic clay, stained (on the thin section) with safranine-b; the early alteration to clay was followed by a congruent dissolution of the amphibole, leaving serrated fragments surrounded by contact voids (v); notice also de ad-sorption of the dye on the weathered parts of the biotite (B). (PPL); **d**) hypo-coating (hc) of methylene blue in conducting voids as result of percolation of stained water through the soil (PPL).

3.2.3.2 Extraction of Carbonates

Removal of finely dispersed calcium carbonate from the groundmass is sometimes necessary to make the masked b-fabric visible (see Section 7.2.4.3); for example a crystallitic b-fabric can hide a striated one.

For the extraction procedure an uncovered thin sections is wholly or partially submersed in 1 N HCl for 3 to 10 min, and then is rinsed with water or ethanol. If necessary, the procedure is repeated. As do-lomite dissolves much more slowly than calcite, the method also gives an indication of the nature of the carbonates present (Wilding and Drees, 1990). Morrás (1973) proposed a treatment with 2% (v/v) HCl to remove calcite from thin sections. During this treatment, part of the thin section may be protected by a layer of silicone grease to preserve part of the original fabric.

3.2.3.3 Extraction of Iron and Manganese Oxyhydrates

Methods for the selective extraction of iron and manganese (hydr) oxides from soil thin sections are discussed by Bullock et al. (1975), Pagliai and Sequi (1982), Arocena et al. (1989 and 1990), and Curmi et al. (1994). In the procedure proposed by Arocena et al. (1990) the uncovered thin section, after being photographed, is placed in a 100-mL centrifuge tube containing sufficient NH_4-oxalate solution to submerge the sample. The tube is shaken in the dark at room temperature for 4 h. The section is then removed and photographed again. The procedure is repeated until the desired amount of Fe or Mn has been removed. This treatment extracts ferrihydrite and the Mn-(hydr)oxides; if longer extraction times are used goethite may also be affected, and magnetite may be attacked (Algoe et al., 2012). After this treatment, the extraction is continued at 75 to 80 °C with a dithionite-citrate-bicarbonate (DCB) solution for 60 min to extract all free Fe.

Curmi et al. (1994) replaced the oxalate procedure by an extraction with citrate–bicarbonate according to Jeanroy et al. (1991). The extraction time is situated between 10 min and 150 h.

For the removal of organically bound Fe Bullock et al. (1975) recommended soaking the uncovered thin section in pyrophosphate for 15 min.

Preparation of solutions:

1. Oxalate: 700 mL of 0.20 M $(NH_4)_2C_2O_4$ and 535 mL of 0.20 M $H_2C_2O_4$; adjust to pH 3.

2. Citrate–bicarbonate: 78.43 g of sodium citrate and 9.82 g of sodium bicarbonate per liter.

3. DCB: 8.4 g of $NaHCO_3$ and 7.52 g of Na-citrate dissolved in 800 mL water; adjust to pH 7.3 with citric acid; adjust to 1 l. Use 50 mL par centrifuge tube; after heating to 75 to 80 °C add 250 to 500 mg of $Na_2S_2O_4$.

4. Pyrophosphate: 0.1 M K-pyrophosphate at pH 10.

Experiments in our laboratory have shown different possible applications of this technique:

1. Gradual stripping of Fe and Mn allows a visualization of the distribution of different types of Fe(hydr)oxides. After each treatment a color micrograph should be made (Plate 3.5).

2. Removal of Fe and Mn (hydr)oxides enhances some microscopic features, such as b-fabrics, and the presence of small clay coatings, masked by Fe (compare Plate 3.5b and e).

3.2.3.4 Bleaching of Humic Substances

Babel (1964) proposed soaking the uncovered thin section a 2.5% (v/v) sodium hypochlorite (NaOCl) solution for 90 min at room temperature to bleach organic matter. This technique is used to (i) distinguish brown and dark humic materials from other (mainly ferruginous) constituents, (ii) make other staining reactions possible and (iii) remove substances masking autofluorescence of tissue components (e.g., cell walls). The reaction seems to penetrate organic matter over the entire thickness of the thin section (Babel, 1964).

3.2.4 Staining and Spot Tests

3.2.4.1 Introduction

Normal optical techniques do not always allow to distinguish chemical or structural related minerals. For instance the different types of feldspars (K, Na, Ca-feldspars) are difficult to identify when twinning is absent, and also simple anhydrous alkaline-earth-carbonates, such as calcite ($CaCO_3$), high-Mg-calcite, ferriferous calcite, aragonite ($CaCO_3$), magnesite ($MgCO_3$), siderite ($FeCO_3$), dolomite ($CaMg(CO_3)_2$) and ankerite ($Ca(Mg;Fe)CO_3$) have very similar optical characteristics in thin sections. In the past, differential staining techniques have been used to distinguish the different types (e.g., Bailey and Stevens, 1960, Houghton, 1980). At present, microprobe analyses, such as EDXRA and WDXRA are commonly used for this purpose. For the feldspars these methods are efficient, although a distinction between the three K-feldspars is not possible. For the carbonates the difficulty remains to distinguish between calcite and aragonite, both having the same chemical composition but a different crystallographic structure. The additional use of micro-XRD or micro-FTIRA is necessary.

3.2.4.2 Differential Staining of Anhydrous Alkali-Earth Carbonates

3.2.4.2.1 Introduction

Although staining is no longer necessary as an identification tool, it may remain useful in the case of complex carbonate materials to understand the spatial, and possibly chronological relation between the different types, such as low- and high-Mg-calcite, ferriferous calcite, and aragonite. Staining can for instance also be used for the quantification of calcium carbonate by image analysis in thin sections (Bui and Mermut, 1989). Staining of the polished surface of the impregnated sample matching with the thin section can be a solution if only covered thin sections are available.

Differential staining techniques have been used for several decades on hand specimens for carbonates (Friedman, 1959, Warne, 1962, Evamy, 1963, Dickson, 1965). Acetate peels, as often used in limestone studies, are rarely used on soil thin sections (Conway and Jenkins, 1977).

3.2.4.2.2. Procedure

Thin sections pose special problems compared to hand specimen: the recommended pretreatment with HCl can completely dissolve the thin layer of carbonates; boiling of the sample is excluded, and so is the use of different solutions after eliminating the earlier obtained stain by etching or polishing. Morrás (1973) described a special technique to overcome these problems. A sufficient quantity of the staining solution is poured into a Petri dish and the uncovered thin section is placed in

Table 3.4. Staining tests for carbonate minerals.

Mineral	Feigl solution	Titanium yellow	Alizarine + ferricyanide	Alizarine	Ferricyanide
Aragonite	black	–	red	red	–
Magnesite	–	yellow	–	–	–
Calcite					
Pure	–	–	red	red	–
High Mg	–	yellow	red	red	–
Low Fe	–	–	violet	red	–
High Fe	–	–	purple	red	blue
Siderite	–	–	purple	–	blue
Dolomite					
Pure	–	–	–	–	–
Fe bearing	–	–	blue	–	blue
Ankerite	–	–	blue	–	blue

the solution with the polished face down, for about 4 to 5 min at room temperature. The solution is agitated from time to time to avoid the formation of air bubbles on the section. Afterward the section is rinsed with distilled water (tap water may produce a staining for Mg or Ca) and air-dried. After observation and photographing, the next staining is performed.

The staining solutions must be used in a specific order. The first treatment with the Feigl solution (see below) will only give a black stain to aragonite. This is followed by a treatment with titanium yellow to stain magnesite and high magnesium calcite. The mixture of alizarine red and potassium ferricyanide is the last to be applied. It will give a red color to calcite, a blue color to siderite, and intermediate colors to Fe–calcite (Plate 3.6a and b). To distinguish calcite from dolomite, a treatment with alizarine red is sufficient; to distinguish between siderite and ferroan calcite, the alizarin red and potassium ferricyanide solutions have to be applied separately. Results of staining are summarized in Table 3.4.

Preparation of solutions:

1. Feigl solution: 1 g $AgSO_4$ added to a solution of 11.8 g $MnSO_4.7H_2O$ in 100 mL water; bring to boiling; filter after cooling; add two drops of diluted NaOH; filter again after 2 h; keep the solution in a cool dark place.

2. Titanium yellow: dissolve 0.2 g titanium yellow in 25 mL methyl alcohol (warm up if necessary); add 15 mL of a 15% NaOH solution.

3. Alizarine red and potassium ferricyanide: dissolve 0.1 g alizarine red S in 100 mL 0.2% HCl solution; dissolve 5 g K - ferricyanide in 1 L of 0.2% HCl solution.

3.2.4.2.3 Application

Apart from the differentiation between carbonate minerals, staining can be used to simply show the presence of minerals. Alizarin staining can for instance be used for the quantification of calcium carbonate by image analysis in thin sections (Bui and Mermut, 1989).

3.2.4.3 Ferruginous Components

Ferruginous components can easily be detected by a simple spot test: treat the section for 5 min with a 5% (v/v) solution of $K_4Fe(CN)_6$ and subsequently with 1 M HCl solution for 30 s. A typical blue color will appear at the places where free iron is present (Babel 1964). The reaction penetrates only a few micrometers deep. If brown or dark stained organic matter is present, it might be useful to bleach it first with sodium hypochlorite (see Section 3.2.3.4.).

3.2.4.4 Manganese Oxides and Hydroxides

Manganese oxides and hydroxides can be recognized by their dull black opaque nature and their effervescence with a 30% (v/v) H_2O_2 solution on uncovered thin sections. Observations under the microscope are best done with oblique incident light; the objective lens should be protected against chemical corrosion, by mounting a cover glass on the objective with a drop of immersion oil.

3.2.4.5 Clay Minerals

Clay can easily be stained in thin sections with dyes such as methylene blue, safranine, and malachite green. Staining can be used to distinguish between fresh and weathered biotite (only the latter absorbs dyes) (Plate 3.6c), as well as to detect kaolinite in fine-grained gibbsite or boehmite in bauxite. A combination of staining with fluoresceine aqueous solution, prior to impregnation, and observation with UV-light has been used by Rassineux et al. (1987) to make microporosity in clay linings visible.

Several tests have been developed for selective staining of clay minerals, but without complete success. Bajwa and Jenkins (1977) experimented with fluorescent dyes on thin sections of pure standard clay minerals using 0.1% (v/v) solutions of acridine yellow, acridine orange, auramicine-o, berberine sulfate, rhodamine-B and thioflavine T. No selective staining was noticed for any of the clay types, but a much stronger adsorption of rhodamine-B by smectite, compared with the other clays, was clear. Staining of soil thin sections with acridine or rhodamine is helpful to visualize the distribution pattern of clay (Plate 3.6c). Best results were obtained by staining the surface of impregnated blocks prior to thin sectioning. According to Deer et al. (1971) kandites are not stained by benzidine, whereas smectites turn light blue. Safranine-Y stains all clay minerals red, and in addition pleochroism is seen with strongly oriented kaolinite. Staining of clay minerals can be a help in their quantification by image analyses.

The property of clay to absorb methylene blue has been proposed as a help for the quantification of conducting porosity in soils (Bouma et al., 1977). In the field, the horizon(s) to be studied are first saturated with water, and later a methylene blue solution is percolated. This will stain the walls of conducting pores only, leaving unconnected voids unstained. In this way, conducting pores can easily be distinguished in thin sections (Plate 3.6d).

4. Elements of Fabric

4.1. INTRODUCTION

Microscopic investigations of soil materials can be divided into two main branches: *compositional studies* and *fabric studies* (Fig. 4.1). The former deals with data such as chemical and/or mineralogical composition or associated characteristics such as color, refractive index, or interference colors. Fabric studies are the major focus of micromorphology and involve two types of data: undirected (scalar) data, and directed (vectoral) data. Scalar data include grain size and shape, shape of boundaries, crystal habit, homogeneity, and distribution patterns of soil components. Vectoral data are related to orientation of components and have been studied only in a few cases in soil micromorphology.

In practice, most micromorphological units are described on the basis of fabric as well as composition. The fabric concepts explained here provide consistent definitions and terms needed to describe both simple and more complex materials, including soils, saprolites, laterites, and calcretes (see also Stoops, 2015a; Stoops and Mees, 2018). Quantitative analyses of micromorphological features are not possible without clear and well-defined concepts (e.g., Krebs et al., 1994).

Nature is in fact a continuum, but to describe and label units, it is to be subdivided in well-defined, closed boxes. It's no wonder that often a partial overlap of units occurs and a given feature appears to fit in several boxes. A detailed, systematic, and imperative key seems

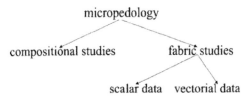

Fig. 4.1. Place of fabric studies in micropedology.

Guidelines for Analysis and Description of Soil and Regolith Thin Sections, Second Edition. Georges Stoops.
© 2021 Soil Science Society of America, Inc. Published 2021 by John Wiley & Sons, Inc.
doi:10.2136/guidelinesforanalysis2

the only solution, but that would take hostage the investigator, and kill any initiative and motivation to continue progress.

4.2. CONCEPTS OF FABRIC

The term *fabric* has been used in geology, soil mechanics, and soil science with different meanings by different authors. In petrography, it has been used both in a very restrictive sense, to describe the orientation of constituents, and also in a very broad sense, comprising "all the textural and structural features of a rock" (Sander, 1948, 1950, 1970). For some authors the terms fabric, texture, and structure are even synonyms.

The concept of soil fabric was introduced in soil micromorphology by Kubiëna (1938, p. 504) for "the arrangements of constituents of the soil in relation to each other". In later publications, he made it clear that he considered the fabric concept not only from a configurational point of view, but also from a functional and genetic one. For Altemüller (1962), soil fabric was the disposition of the soil constituents (solid, liquid, and gaseous), the building blocks of the soil, in a morphological, functional, and genetic sense. Brewer and Sleeman (1960) restricted the concept of soil fabric to the spatial arrangement of soil constituents.

Bullock et al. (1985) proposed an open fabric concept, based mainly on Altemüller (1962).

> *Soil fabric* is the total organization of a soil, expressed by the spatial arrangements of the soil constituents (solid, liquid and gaseous), their shape, size and frequency, considered from a configurational, functional, and genetic viewpoint. (Bullock et al., 1985, modified).

At present, it is generally not possible to distinguish between the liquid and the gaseous phases, although successful efforts have been made in the case of petrol-bearing sediments to distinguish in thin sections between solid (sand), liquid (petrol), and gaseous components (Bell et al., 2013).

In practice, it is impossible to express the total fabric of a soil material because of its complex nature, and only selected aspects of the fabric can be described. The fabric elements described should be chosen in such a way that they be causally related to the studied characteristics. Furthermore, the fabric one observes also depends on the method and scale of observation, and the experience (degree of understanding) of the observer. The *observed soil fabric* deals with the arrangement, size, shape, and frequency of the different fabric units in soil material, observed at a given scale of observation, and with a given method. The *integrated soil fabric* is formed by the combination of several observed soil fabrics, studied at different levels of magnification and using different methods (e.g., PPL, XPL, UVF, CL) (Plate 3.3).

BACKGROUND - Kubiëna (1938, p. 504) defined soil fabric as "the arrangement of the constituents of a soil in relation to each other". The lowest level is the elementary fabric that is "the arrangement of the constituents of lowest order in the soil in relation to each other" and corresponds more or less to the coarse/fine (c/f) related distribution of the groundmass (see Section 7.2.2). According to Brewer and Sleeman (1960) and Brewer (1964a) soil fabric describes the spatial arrangement (i.e., orientation and distribution patterns) of solid particles and associated voids, and corresponds as such to one of the elements of fabric considered by Bullock et al. (1985). Arrangement, size, and shape of solid particles and associated voids combined define "soil structure" according to Brewer (1964a). This definition of structure thus partly corresponds more or less with fabric, as defined by Bullock et al. (1985), and is not related to structure and pedality as used in the field.

Fabric must be considered to be of infinite extent. Therefore, neither the bounding surfaces nor the shape of the body are part of its fabric (e.g., as in geology where a material is called conglomeratic, irrespective of whether it is observed in a quarry or as a polished pillar in a church). In fabric analyses, absolute size of the units is not considered.

Fabric can be described only on the basis of heterogeneity. A completely homogeneous body, a blue sky for example, displays no fabric. Homogeneity depends on the scale of observation: a body may seem homogeneous at one scale, but heterogeneous when studied at higher magnification. Strict homogeneity is not achieved in nature because material itself is basically discontinuous (e.g., a crystal lattice), therefore only *statistically homogeneous zones* exist.

Basically the degree of homogeneity or heterogeneity (the geneity according to Sander, 1950) can be measured in two ways: (i) by comparing the variance in a number of equally-sized samples, or (ii) by determining the size of randomly selected test samples which seem identical at the scale of observation. The more a material is homogeneous, the smaller the test sample will be. For example in a coarse conglomerate, the test sample must be several centimeters in size, in a sandstone several millimeters, and in a clay only a few micrometers (see also Fig. 4.2).

The second method is recommended because the first method requires numerical data, generally not available or only available at the end of a micromorphological study. For a given scale of observation, the sample size (either expressed in area or, more conveniently, by its diameter), necessary to reduce variability to zero, or to a preset limit, determines the geneity. VandenBygaart and Protz (1999) defined this as the *representative elementary area* (REA). This is the smallest area where the value of a property measured on three successive areas of measurement does not change by 10% relative to the next greater area of measurement. It was found that the REA corresponds generally to an area four to six times the diameter of the largest component within the materials under study.

Homogeneity can be total, if all characteristics of the sample are considered, or it can be partial, as when a single given characteristic is examined. A soil may be homogeneous with respect to one feature, whereas for another feature it may be heterogeneous. For instance the distribution of quartz grains may be homogeneous in a soil, but that of illuviation features may be heterogeneous. Geneity is also not only funcüon of scale, but also function of the observation method used. An opaque area may seem homogeneous in plane polarized light (PPL) but it can be clearly heterogeneous in oblique incident light (OIL) (Plate 3.3d, through f).

A fabric unit is a finite, three-dimensional unit delimited by natural boundaries, statistically homogeneous on the scale under consideration, and that can be distinguished from other fabric units by the methods of study applied and at the scale of observation used (Stoops, 1978a; Bullock et al., 1985).

At each level of observation, the homogeneous zones are fabric units (Fig. 4.2). Examples are the clay matrix, quartz grains, Fe-(hydr) oxide nodules and clay coatings. According to the definition of fabric given above, voids are to be considered as fabric units.

The simplest fabric units are the *basic components*, constituting the building blocks of the micromass, groundmass, and pedofeatures (see Chapter 6). They can be inherited (quartz grains, feldspar grains, rock fragments) or pedogenic (gibbsite, gypsum, clay). Thus a rock fragment or a plant tissue fragment is a basic component notwithstanding a highly differentiated internal fabric. Clay and colloids are considered in thin sections as basic components because their individual particles, if any, are below the resolution of the petrographic microscope.

The criterion of homogeneity used in the definition of fabric units should be considered in a broad sense, taking into account the possibility of partial homogeneity, such as the variability tolerance within the statistical homogeneity.

When several fabric units have the same characteristics (at the scale of observation and with respect to the applied criteria), they can be considered as a partial fabric (Teilgefüge, according to Sander, 1948). Imagine, for instance, a thin section with several similar Fe-hydroxide nodules. Each nodule is a fabric unit, but all nodules together form a partial fabric (Fig. 4.2). In the same way, all quartz grains of the groundmass together form a partial fabric. The voids of a soil may also be considered as a partial fabric, but they can be split up in different partial fabrics, according to morphology (e.g., fissures, channels, vughs), or according to function (e.g., conducting pores vs. closed pores) for example. The concept was applied for the first time in micropedology by Kubiëna (1956) to distinguish two partial fabrics in European Luvisols: the groundmass (Braunerde Teilgefüge) and the illuviation coatings (Braunlehm Teilgefüge).

"A partial fabric of a soil material comprises all fabric units, whether interconnected or not, which are on a given scale identical with respect to the criteria considered" (Stoops, 1978a; Bullock et al., 1985).

Partial fabrics can also group materials on a less configurational and more abstract or process-oriented basis. For example, the parental

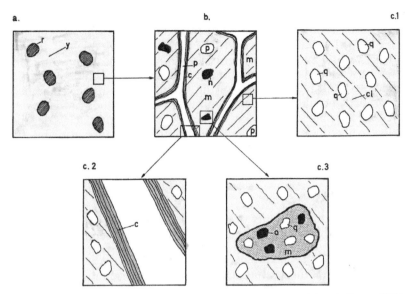

Fig. 4.2. Homogeneity, heterogeneity, fabric units, and partial fabrics – a look at their scale (Stoops, 1978a).
(a) Mottled clay: seven homogeneous zones, or fabric units, can be distinguished with the naked eye: six reddish mottles (r) in a yellowish matrix (y); the six red mottles together form one partial fabric, the yellow matrix another one; **(b)** at low magnification the apparently homogeneous matrix shows itself to be composed of four homogeneous partial fabric: a matrix (m), Fe hydroxide nodules (n), clay coatings (c) and pores (p); **(c)** at higher magnification these apparently homogeneous partial fabric are heterogeneous, being constituted in turn of several homogeneous partial fabrics: (c.1) quartz grains (q) in a clay matrix (cl), where each quartz grain is a fabric unit, all quartz grains together form a partial fabric; (c.2) the clay coating (c) consists of an alternation of Fe-rich and Fe-poor zones where each type forms a partial fabric; (c.3) the Fe hydroxide nodules consist of three homogeneous components or partial fabrics: a dark brown matrix (m) with opaque manganese micronodules (o) and quartz grains (q).

partial fabric (grouping all constituents taken from the parent material), the inherited partial fabric (grouping all features inherited from a former pedogenesis), the pedogenic partial fabric (grouping all features formed by the present pedogenesis). Such groupings are of course no longer purely descriptive, but involve interpretation. Other examples could be the redoximorphic partial fabric, grouping all features related to oxidation–reduction. Within a partial fabric, a preset degree of variability has to be taken into consideration (see Section 4.4).

4.3. ELEMENTS OF FABRIC

4.3.1. Introduction

The characteristics that are used to describe the fabric of a soil are called the elements of fabric. The most important elements of fabric are the spatial distribution and orientation, size and sorting, and shape. Color is also treated in this chapter as a descriptive criterion, although it is not a part of the fabric as such. Some fundamental criteria, such as size and shape, used in description are common to a number of different constituents in thin sections. To avoid repetition under several headings, they are treated as a whole in this chapter.

Three levels of increasing accuracy of descriptions are considered: (i) visual description– the component is described from visual observation only, (ii) visual comparison– the particular component is compared with a set of reference components, and (iii) direct measurement. General criteria used for visual description or measurement include pattern, size, abundance, and relative area, sorting, color, shape, surface roughness or smoothness, boundary, and variability. For each of the criteria described here, it is possible to adopt any of the above three approaches depending on the required level of detail and the resources and time available.

4.3.2. Patterns

4.3.2.1. Introduction

One of the main elements of fabric is the spatial arrangement of the fabric units, which is called their *pattern*. Arrangement is characterized according to the orientation and distribution pattern of the fabric units.

A *pattern* is the spatial arrangement of fabric units.

The definition given above is more general than the one given by Bullock et al. (1985). Brewer and Sleeman (1960) and Brewer (1964a) gave a comprehensive discussion of arrangements.

Two main types of patterns can be considered: distribution patterns and orientation patterns. For each type, *basic, referred,* and *related patterns* can be distinguished (Fig. 4.3.). Distribution patterns result from a relative quantitative increase of the partial fabric under consideration. For example, a decrease in the distance between quartz grains in an area gives rise to a clustered distribution. If the decrease is concentrated in the neighborhood of a reference feature (e.g., a channel) then a referred distribution pattern results. In the case of elongated fabric units it is sometimes confusing whether a distribution or an orientation pattern is observed.

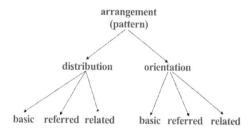

Fig. 4.3. Elements of spatial arrangement of fabric units.

4.3.2.2. Basic Patterns

Basic distribution or orientation patterns are the distribution or orientation of fabric units of the same type with regard to each other (Brewer, 1964a, modified).

4.3.2.2.1. Basic Distribution Patterns

The following types of basic distribution patterns can be distinguished (Fig. 4.4):

Random: the distribution of the individuals is statistically random.

Clustered: individuals occur in groups; the distance between the individuals in a group is smaller than the distance between the groups.

Linear: individuals are distributed according to straight lines.

Banded: individuals are concentrated in bands; the distance between the individuals in the bands is smaller than the distance between the bands; bands may be continuous or discontinuous (Plate 4.1a; Plate 8.22a and b). Sub-types include: *straight, sinuous, tubular and bow-like*. Banded basic distribution patterns of sand, silt or clay are frequently observed in the lower part of soils developed on alluvium.

Circular and **oval**: individuals are distributed according to circles or ovals.

Fan-like: individuals are spread according to a fan-like pattern.

Interlaced: the individuals are interlaced with each other.

Basic distribution patterns can be evaluated quantitatively using statistical methods (Arocena and Ackerman, 1998).

4.3.2.2.2. Basic Orientation Patterns

Following Brewer (1964a), Bullock et al. (1985) distinguished two types of basic orientation- *random* and *parallel*- with four degrees of parallel basic orientation. This system proved to be incomplete because bimodal orientations were not considered, and the distinction between the quantity of oriented particles, and their angle of parallelism was not clear. Therefore, three types of basic orientation are proposed: *unimodal parallel, bimodal parallel* (e.g., in bistrial b-fabrics, see Section 7.2.4.2) and *random*.

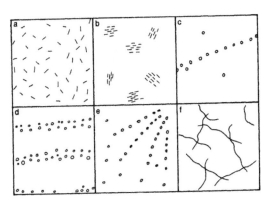

Fig. 4.4. Basic distribution patterns. **(a)** random, **(b)** clustered (note parallel basic orientation in clusters), **(c)** linear, **(d)** banded, **(e)** fan-like, and **(f)** interlaced.

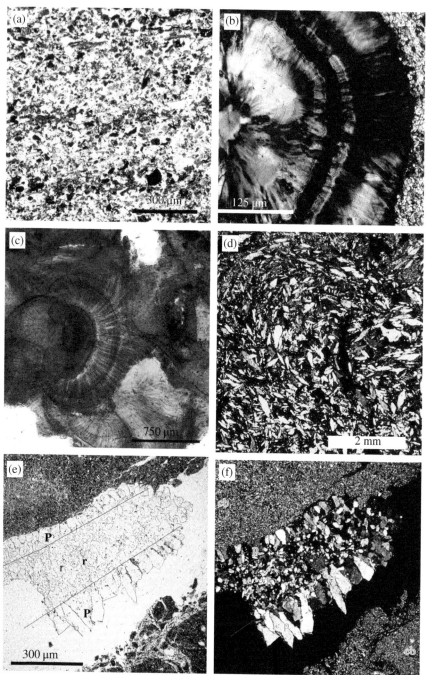

Plate 4.1. Distribution and orientation patterns. a) banded distribution of sand grains and fine (dark) organic fragments with weak parallel orientation (both basic and referred) of elongated organic particles in alluvium (PPL); **b)** detrital grain of chalcedony with concentric distribution and radial orientation pattern of the quartz fibres (XPL); **c)** concentric distribution and radial orientation of goethite fibres in bog ore (PPL); **d)** crescent pattern of lenticular gypsum crystals in a gypsic horizon; passage feature (XPL); **e)** part of compound gypsum coating on a plane with parallel distribution of three zones: in the middle zone the crystals are oriented at random (r), in the inner and outer zones they are oriented perpendicular (p) to the wall (so called palisade fabric) (PPL). **f)** same, calcitic crystallitic b-fabric (cb) in the groundmass (XPL).

The description of the degree of orientation should take into account the *degree of parallelism*, expressed as the angle of mutual orientation, and the quantity of oriented particles expressed as a percentage of the total amount of particles belonging to the same partial fabric. The following terms are recommended to describe the degree of orientation:

Degree of Parallelism

Small angle: individuals oriented with their principal axes within 30° of each other (i.e., within 15° of a line representing the mean orientation).

Medium angle: individuals oriented with their principal axes within 60° of each other (i.e., within 30° of a line representing the mean orientation).

Broad angle: individuals oriented with their principal axes within 90° of each other (i.e., within 45° of a line representing the mean orientation).

Quantity of oriented particles:

Strongly-expressed: > 66% of the individuals of a same partial fabric are oriented.

Moderately expressed: 66% to 33% of the individuals are oriented.

Weakly expressed: < 33% of the individuals are oriented.

Both terms can be combined, eventually with the indication of a bimodal trend. For example, a strongly-expressed medium angle of orientation combined with a weakly expressed small angle of orientation in a bimodal pattern where > 66% of the individuals are oriented with their principal axes within 60° of each other, < than 33% show a different orientation within 30° of each other.

Basic orientations can be graphically illustrated using orientation diagrams (Hill, 1970 and 1981). Because only the orientation, not the direction is measured, it is unnecessary to use a 360° diagram, as done in climatology (so called wind roses) but a 180° diagram. Three dimensional orientations of single mineral grains, such as quartz grains, can be plotted in stereodiagrams. This has been done in micromorphology for instance for morainic or fluviatile materials (Korina and Faustova, 1964; Sen and Mukherjee, 1972) (see also Stoops and Mees, 2018).

4.3.2.2.3. Orientation Patterns of Clay Particles

A special case is the expression of the basic orientation of clay particles. Clay particles are mainly phyllosilicates (monoclinic or triclinic) and, therefore, optically anisotropic. They are biaxial and length slow in sections parallel to the (crystallographic) c-axis, showing their foliated fabric. Theoretically, their interference colors should be visible when observed between crossed polarizers, but because of their small dimensions (< 2 µm) compared with the thickness of a thin section (20–30 µm), overlapping occurs and no individual grains can be observed in transmitted light. Interference effects are counterbalanced by other particles if their orientation is random (so-called "card-house

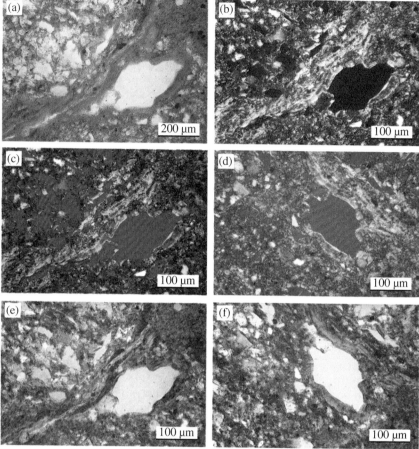

Plate 4.2. Orientation of clay particles. a) coating of fine clay with continuous orientation (PPL); **b)** same; notice interference colors of the clay up to red first order and extinction bands (XPL); **c)** same as b, but with λ-retardation plate (so called gypsum plate) inserted, with nγ oriented NE-SW; notice the higher interference colors, up to blue first order (XPLλ); **d)** same as c, but thin section rotated over 90°; notice that interference colors get lower than in b for clay oriented NW-SE (XPL,λ); **e)** same as b, but with ¼ λ retardation plate (so called mica plate) inserted, notice higher interference colors compared to b (XPL, ¼ λ); **f)** same as e, but thin section rotated over 90°; notice lower interference colors (XPL, ¼ λ).

Plate 4.3. Orientation of clay particles. a) striated orientation due to parallel orientation of elongated clay domains (striae) at the side; at the right side locally small equidimensional domains forming a speckled pattern. (XPL); **b)** same, with λ retardation (nγ oriented NE-SW) resulting in higher interference colors of the NE-SW oriented domains. (XPLλ).

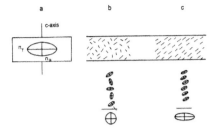

Fig. 4.5. Optical effect of random and parallel oriented clay in thin section. (**a**) orientation of the slow (nγ) and the fast (nα) ray (section through the indicatrix) in a phyllo silicate plate; (**b**) random oriented clay particles: interaction between slow and fast rays leads to a statistical isotropy; (**c**) parallel oriented clay particles: slow and fast rays of the different particles enhance each other, leading to a pronounced orientation birefringence.

arrangement") (Fig. 4.5.a). In such a case, the clay-mass seems isotropic (*statistical isotropy*) and no light will reach the ocular when the polarizers are crossed (Stephen, 1960).

If the clay particles are oriented parallel to each other, however, they will act as a single crystal, because birefringence effects of the individual particles will be mutually enhanced by that of the other clay particles with the same orientation (Fig. 4.5.b). In most cases, clay particles form small aggregates (~ 20–30 μm) of oriented plates, the so-called *domains* (Aylmore and Quirk, 1959), or *pseudocrystals* (Russian terminology). In electron microscopy, this appears as a face-to-face arrangement. Optical anisotropic zones caused by parallel arrangement of clay particles were mentioned already by Mitchell (1956), Minashina (1958), Morgenstern and Tchalenko (1967a) and Stephen (1960). If a large number of particles have perfect parallel orientation (as in the case of illuvial fine clay coatings), extinction figures are clear (e.g., extinction lines in curved clay features) (Plate 3.2a, Plate 8.5b, Plate 8.6b). A weak pleochroism is observed if the clay is stained (e.g., by Fe-oxihydrates or some dyes). The best continuous orientation is observed in the case of fine (< 0.2 μm) clay. As the grain size of the clay increases, orientation generally becomes less perfect, and therefore the extinction lines become less sharp. In tangential sections through clay coatings or infillings, the central part often seems isotropic because clay flakes are observed parallel to their optical axis. This is the best position to determine the axial figures in conoscopic light. Orientation of clay particles may be due to sedimentation (clay coatings, crusts), compaction (Mitchell, 1956, Moon, 1972) or shearing (Morgenstern and Tchalenko, 1967b) (e.g., slickensides). A more complete discussion is given in Stoops and Mees (2018).

Because for all phyllosilicates the largest refraction index is oriented perpendicular to the c-axis, all sections, apart from those parallel to (001), are length slow. This can help to determine the orientation of the clay. For instance in illuvial clay coatings the particles are oriented parallel to the void walls, and the coating as a whole will show higher interference colors (generally blue) when the λ retardation plate is inserted (Plate 4.2, Plate 8.6a through c) with n_γ parallel to the coating. In coatings of newly formed clay the particles are generally oriented perpendicular to the walls. The coating as a whole behaves as a "length fast" elongated body and will show lower interference colors (generally yellow) when the retardation plate is inserted (Plate 8.6d through f).

Interference colors for groups of elongated crystallites that are oriented parallel to one another may be influenced by the refractive index of the medium between the individual particles (the so-called *form birefringence*). In this way, even parallel arrangements of isotropic rods embedded in an isotropic medium with a different refractive index can behave as an anisotropic body, because the sum of the refractive indices parallel to the length of the rods and that perpendicular to it will be different. This can be for instance the case in bone fibrils (see 6.2.4.5 and Plate 6.10). Much more research on this topic is needed.

The degree of orientation of the domains is expressed by the following terms:

Continuous orientation: all clay particles have a parallel orientation; the feature exhibits extinction lines or extinguishes as a unit (Plate 3.2a. Plate 4.2, Plate 8.5a and b, Plate 8.6a through c).

Striated orientation: small elongated clay domains (striae) show a linear or banded arrangement (Plate 4.3).

Flecked orientation: equidimensional clay domains are randomly oriented.

The orientation of the clay domains can be graphically illustrated in orientation diagrams (e.g., Hill, 1970 and 1981) (see also Stoops and Mees, 2018).

4.3.2.3. Referred Patterns

Referred distribution or orientation patterns are the distribution or orientation of like fabric units with respect to a reference (Bullock et al., 1985, modified).

Examples of reference features are grain or void surfaces. The reference features needs not necessarily be visible in the thin section, for example the soil surface.

4.3.2.3.1. Referred Distribution Pattern

The following types of referred distribution pattern can be distinguished (Fig. 4.6.):

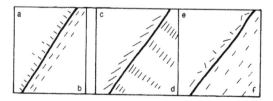

Fig. 4.6. Referred distribution and orientation patterns of fabric units: (**a**) parallel distribution, perpendicular orientation, (**b**) parallel distribution and orientation; (**c**) parallel distribution, oblique orientation; (**d**) perpendicular distribution, parallel orientation; (**e**) parallel distribution, random orientation; (**f**) random distribution, parallel orientation.

Random or unreferred: the pattern of distribution is unrelated to the reference feature.

Perpendicular: the individuals are distributed perpendicular to the reference feature.

Parallel: the individuals are distributed parallel to the reference feature (Plate 4.1e and f). Coated is a particular type of parallel distribution pattern, although not all coatings have also a parallel internal distribution pattern.

Inclined: the individuals are distributed at an angle with the reference feature.

Radial: the individuals are grouped along radiating lines.

Concentric: the individuals are grouped along approximately concentric lines or surfaces (Plates 4.1b and c, Plate 8.15e and f)

Bow-like or **crescent-like:** semielliptical arrangements of soil constituents between subparallel walls (Plate 4.1d, Plate 8.4, Plate 8.10c and d).

4.3.2.3.2. Referred Orientation Patterns

Referred orientation patterns can be described in terms similar to those used for referred distribution patterns (see above) Terms include: *unreferred, perpendicular* (Plate 4.1e and f), *parallel, inclined, radial* (Plate 4.1b and c, Plate 8.13a through d), *concentric and bow-like* (Fig. 4.7) (Plate 4.1d, Plate 8.2a and b).
  Referred distribution and orientation patterns can also be expressed in a more absolute way, for example (sub)horizontal, (sub)vertical.

4.3.2.4. Related Distribution Patterns

4.3.2.4.1. General

> *Related distribution or orientation patterns* are the distribution or orientation of like fabric units in relation to fabric units of another type (Bullock et al., 1985, modified).

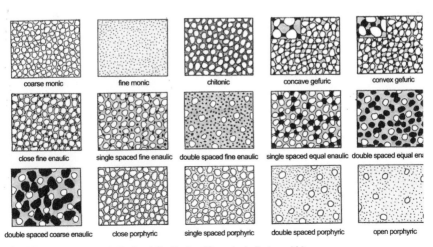

Fig. 4.7. Different types of c/f related distribution (blue color indicates voids).

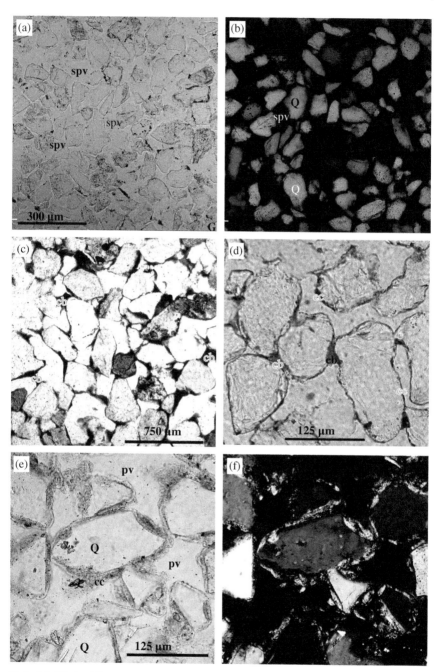

Plate 4.4. C/f related distribution patterns. a) coarse monic: quartz grains and simple packing voids; single grain microstructure (PPL); **b**) same: the quartz grains (Q) seem to be floating and show only few points of contact (spv: simple packing voids); single grain microstructure (XPL); **c**) concave gefuric; bridged grain microstructure (PPL); **d**) gefuric (cb: clay bridge) and thin chitonic (PPL); **e**) chitonic: free grain clay coatings (cc), quartz grains (Q) and packing voids (pv); pellicular grain microstructure (PPL); **f**) same (XPL).

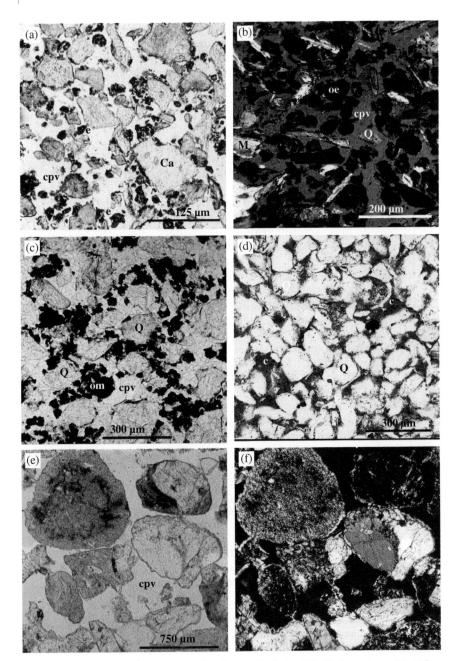

Plate 4.5. C/f related distribution patterns. a) close fine enaulic c/f related distribution pattern or intergrain microaggregate microstructure, formed by excrements between angular calcite grains (Ca), complex packing voids (cpv) (PPL); **b)** double spaced fine, ranging to equal, enaulic c/f related distribution pattern, complex packing voids (cpv). (oe: organic excrements, Q: quartz, M: muscovite) (partly XPL). **c)** close fine enaulic c/f related distribution pattern or intergrain microaggregate microstructure; complex packing voids (cpv). (om: organic microaggregates, Q quartz). (PPL); **d)** close porphyric c/f related distribution: dense packing of coarse quartz grains (Q) and infilling of all packing voids by clay (c) (PPL); **e)** coarse monic c/f related distribution of poorly sorted rock fragments; single grain microstructure; simple packing voids (PPL); **f)** same as e; a thin clay coating becomes visible by its interference colors, indicating that the c/f related distribution pattern is in fact mono-chitonic and therefore voids correspond to compound packing voids (XPL).

The same general terminology used for referred arrangements may be used for related arrangements: *unrelated, parallel, perpendicular, and inclined*. Although the related distribution pattern can be inferred from the referred distribution pattern of the two groups of individuals, it has become customary to assign a name to significantly different patterns.

4.3.2.4.2. The c/f related distribution pattern

BACKGROUND - Kubiëna (1938) called the related distribution between plasma and skeleton grains the elementary fabric and introduced terms to describe different types. They are however not purely distribution patterns, as also the nature of the skeleton grains and plasma are taken into account. Brewer (1964a) proposed four terms for the main types of related distribution of plasma and skeleton grains in the s-matrix (see also Section 7.2.2). Stoops and Jongerius (1975, 1977) established a system for the description of the related distribution of small fabric units in relation to larger fabric units, namely the c/f-related distribution (c/f denotes coarse vs. fine). They proposed a floating size limit between coarse and fine, to be determined for each individual case, and they placed no limitations on the nature or complexity of the particles involved. As such, their system is not restricted to the fabric of a groundmass, but also applies to the description of fabrics of higher order (e.g., arrangement of pedofeatures or of aggregates). Later, Brewer and Pawluk (1975) extended Brewer's four original patterns to include relationships between matrix and framework members (i.e., fine and coarse basic constituents) (see Section 7.2.2). Eswaran and Baños (1976) proposed a system of related distribution between colloidal-sized, silt-sized and sand-sized material, but this was only used in few publications in the past.

> *The c/f related distribution* expresses the distribution of individual fabric units in relation to smaller fabric units and associated voids (Stoops and Jongerius, 1975).

The main types of c/f-related distribution patterns and their relationship are illustrated in Fig. 4.7 and 4.8. A short discussion is given below. Original definitions have been adapted to allow the introduction of subtypes.

Monic: only fabric units larger (coarse monic) or smaller (fine monic) than a given size limit, and associated interstitial voids are present. Coarse monic can be seen in sands or gravels (Plates 4.4a and b, Plate 4.5e), and fine monic in clays, but in the latter case associated voids will not be visible with an optical microscope.

Gefuric: braces of smaller fabric units (e.g., in incipient spodic horizons) link the large fabric units. According to the morphology of the bridges concave gefuric and convex gefuric are distinguished (Plate 4.4c and d).

Chitonic: a cover of smaller units surrounds the larger fabric units (e.g., in sandy argillic or spodic horizons) (Plate 4.4e and f, Plate 4.5f, Plate 6.15e and f).

Enaulic: the smaller units form aggregates, which occur in the interstitial spaces between the larger units (e.g., in loose spodic horizons and some A-horizons). The aggregates should not fill the complete void space. Stoops et al. (2001) distinguished subtypes based on:

(i) the relative distance between the coarse grains:

Close enaulic: the coarser units have points of contact (not necessarily visible in thin sections, see Section 3.1.2) (Plate 4.5a and c).

Single spaced enaulic: the distance between the coarser units is less than their mean diameter.

Double spaced enaulic: the distance is one to two times the mean diameter (Plate 4.5b).

Open enaulic: the distance between the coarser units is more than twice their mean diameter.

(ii) the relative size of the coarser grains and the aggregates:

Fine enaulic: the aggregates of fine material are considerably smaller than the coarser units (Plate 4.5a and c).

Equal enaulic: the aggregates have approximately the same size as the coarser material (Plate 4.5b, Plate 7.1c and d).

Coarse enaulic: the grains of coarser material are considerable smaller than the aggregates.

(iii) less important, the shape of the aggregates: *rounded smooth, rounded rough, subangular smooth, subangular rough enaulic*.

A combination of the terms of the subtypes allows precise description, for example "close fine rounded enaulic". Some combinations are impossible, such as close coarse enaulic. Most combinations have a genetic meaning, and are related to the gradual variation between grain structures and granular structures (see Section 5.4.1 and Fig. 5.7).

Porphyric: the larger fabric units occur in a dense groundmass of smaller units. Several subtypes are distinguished based on the relative distance between the coarser units (compare enaulic): *close* (Plate 4.5d, Plate 8.9d), *single spaced* (Plate 6.5a), *double-spaced* (Plate 8.1a and b) and *open* porphyric, (Plate 7.2a through f, Plate 7.3).

The five basic types can be represented on a trigonal bipyramid, in which the corners in the equatorial plane are occupied by chitonic, close enaulic, and gefuric, and the top and bottom corners, respectively, by coarse and fine monic (Fig. 4.8). On the edges, intergrades between two types are represented, on the faces intergrades between three types and inside the pyramid intergrades between four types. The porphyric types are inside the pyramid inserted in the lower part. The top of this inserted pyramid corresponds to a close porphyric type, toward the bottom it becomes more and more open porphyric and will finally become

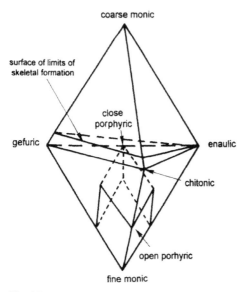

Fig. 4.8. Relation between the five basic types of c/f related distribution, the surface of limits of skeletal function and the surface of limits of porphyric types (Stoops and Jongerius, 1975, modified).

fine monic. Single spaced to open enaulic c/f related distribution patterns do not fit in this scheme. In Fig. 4.8 an additional surface cuts the bipyramid in two unequal parts; it is the meeting place of all limits of the skeletal function (provided the coarser units are hard in dry and wet conditions). Above the surface, the coarser units have a skeletal function where they touch and support one another, and soils are not sensitive to compaction. Beneath the surface, no skeletal function exists. It is also the limit of vertic behavior where wetting may produce a swelling of the soil material. According to E. Braudeau (personal communication) this limit is situated at a silt plus clay content of about 10%. In sediments, grain-supported fabrics occur above the surface, matrix supported ones beneath it.

Intergrades are named by joining the two or three terms, the most dominant being mentioned last. From the first term, the suffix -ic and the last consonant is dropped, for example *chito-gefuric, mono-chitonic* (Plate 4.5f). If both types are present in approximately equal amounts, the terms are written out fully: *gefuric-enaulic*. The c/f-related distribution in a soil thin section is frequently not homogeneous, and two or more types occur besides each other in discrete zones. In that case, the terms are joined by the word "and", for example, "chitonic and gefuric".

FitzPatric (1984) objected to the concept of c/f related distributions because in many soils grain sizes forms a continuum, whereas the c/f concept suggests a bimodal size distribution. A study of many micromorphological descriptions has shown that only in a relatively small number of soils does this problem arise (e.g., in soils on glacial tills). In most cases, there is a striking difference between coarser and finer components (e.g., in soils on sorted sediments, and in most tropical soils), and the description of the c/f related distribution has proven to be most useful.

BACKGROUND - Stoops and Jongerius (1975) present a complex terminology to indicate, apart from the type of c/f related distribution, the size classes and the material. In their terminology following symbols were used for grain size: psef- (gravel), psam- (sand), sil- (silt) and pel- (clay). To indicate the material, symbols were coined by adding the suffix -o to the name of the material or part of it. For example, chitonic quartz-opsam/argiopel points to a material consisting of sand-sized

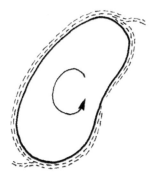
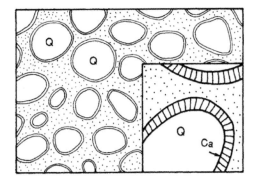

Fig. 4.9. Turbate or galaxy fabric: tails of oriented fine material are formed as a result of rotational movements of the grain.

Fig. 4.10. Different levels of arrangement of fabric units. Looking to the general picture, a random basic distribution of quartz grains and a chitonic c/f related distribution of coarse quartz (Q) and fine calcite (Ca) crystals is noticed. The detail shows clearly that the calcite grains forming the coating have a parallel basic orientation. Taking the surface of the quartz grains as a reference, the distribution of the calcite crystals is parallel, and their orientation perpendicular.

quartz grains with coatings of clay; enaulic calcopsam/humipel points to a material consisting of aggregates of clay-sized humic substances occurring between sand-sized calcite grains, as observed in some Rendolls.

Preferred epitaxial growth of one mineral on another is a specific case of related distribution, frequently considered as a related orientation pattern. A specific related distribution pattern is found in *turbate structures* or *galaxy structures*, consisting of a harder-coarse nucleus, surrounded by finer grains spiraling the nucleus, while tips of the spiral arms extend out from the nucleus (Fig. 4.9) (van der Meer, 1997)

It is not always possible to describe patterns in terms of either basic, referred, or related orientations or distributions. In many cases the descriptions have to be made on different levels (Fig. 4. 10). In some cases combinations of orientation and distribution patterns have to be considered. For instance, most strial b-fabrics (Section 7.2.4.2) are based on the distribution pattern of orientation patterns.

4.3.3. Size

Although absolute size is not an element of fabric, relative size does represent one. As already explained in Section 3.1.2, size of objects > 20 μm measured in thin sections usually does not correspond to their real size. Smaller particles often cannot be recognized because of overlap in thin sections (see also Section 3.1.4). In the case of units that are not equidimensional, the largest and smallest diameter should be recorded. For features such as crusts, coatings, or fissures, the thickness can be described.

Size can be measured directly, using a micrometer eyepiece or by image analysis, or indirectly on photographs (e.g., with a ruler). Exact size measurements are usually not required for a general description, so visual estimates usually suffice. A chart (Fig. 4.11) has been developed by Stoops (1981) as a student aid to estimate grain sizes in thin sections.

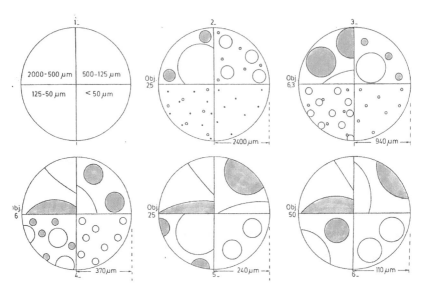

Fig. 4.11. Visual aid for estimating sizes in thin sections (Stoops 1981). Each circle is divided into four quadrants representing the minimum and maximum size of grains belonging to coarse sand, medium sand, fine sand and silt respectively, at different magnifications (objectives 2.5, 6.3, 16, 25 and 50) for a Leitz Orthoplan. The bar scale gives the size of the radius of the field of view. For each combination of lenses (ocular, objective, tube) a new diagram has to be drawn.

The chart shows a series of large circles, representing the field of view at increasing magnifications. Each circle is divided into four quadrants, similar to the quadrants produced by the cross hair of the ocular in a microscope. The minimum and maximum size of grains belonging to a given size fraction, when cut along their largest diameter, are illustrated. In the example of Fig. 4.11 the following size fractions are represented clockwise: coarse sand (2000- 500 μm), medium sand (500-125 μm), fine sand (125-50 μm) and silt (< 50 μm). Of course it is possible to make more subdivisions, but this would reduce the clarity of the chart. The radius of the field of view is indicated as a bar scale. The chart is made as follows: for each combination of objective and ocular the diameter of the field of view is measured accurately with a stage micrometer and recorded as a bar scale. The diameters corresponding to the selected size limits (e.g., 2000, 500, 125, 50 μm) are calculated and represented by circles in the quadrants. An individual chart should be made for each microscope, as the final magnification depends on the combination of tube factor, oculars and objectives used. A similar chart can be prepared for the PC screen if this is used for observations and analysis.

The most appropriate size classes for mineral grains are those used to report particle size distribution (e.g,. the IUSS or USDA-system) so that a continuity with field description is achieved. The descriptive terms given in Table 4.1 are recommended. For units with a size > 2 mm, the (approximate) size should be indicated in millimeters or centimeters. As limits may be other for different classification systems, it is absolutely necessary to specify the particular size classes used.

Table 4.1. Recommended terms to describe the sizes of fabric units.

Class	Size limits (in μm)
Fine clay	< 0.2
Clay	< 2
Silt	2–20 (or 2–50 or 2–63)
Very fine sand	20–100 (or 50–100 or 63–100)
Fine sand	100–200
Medium sand	200–500
Coarse sand	500–1,000
Very coarse sand	1,000–2,000
Fine gravel	> 2,000

The terms *cryptocrystalline* (individual crystals not visible with the optical microscope and too small to show polarization), *microcrystalline* (crystals identifiable only with a microscope) and *macrocrystalline* (crystals visible with hand lens or naked eye) have been commonly used in petrography to denote the size range of crystals in minerals aggregates or rock fragments. Such terms can also be applied conveniently to such constituents in soils. In sedimentology the terms *micrite* and *sparite* are used to indicate fine (< 5 μm) and coarse (> 5μm) calcite, respectively. In sedimentology they have a genetic meaning, not taken into consideration in micromorphology.

4.3.4. Sorting

Sorting is defined statistically as the amount of variation in particle diameters found in a sample. The following classes can be used for rough visual estimates:

Perfectly sorted: normally only one size fraction is present.

Well sorted: 5 to 10% of sizes other than the dominant size fraction.

Moderately sorted: 10 to 30% of sizes other than the dominant size fraction.

Poorly sorted: > 30% of sizes is other than the dominant size fraction.

Unsorted: the particles are of a variety of sizes with no fractions appearing more dominant than others.

The apparent sorting of particles > 20 μm in thin sections deviates from its actual sorting due to the fact that only a few grains are cut along their largest diameter (see Section 3.1.2).

4.3.5. Abundance

The abundance of fabric units can be expressed in different ways, such as the number of units per surface unit (e.g., number of quartz grains per square centimeter), or the surface percentage occupied by a partial fabric. The use of descriptive terms such as rare, common, or frequent can be misleading because areas occupied by different fabric units may be quite different. For example, 2% of celestite in an arid soil is very high, whereas a same amount of gypsum in a similar soil is very low. Therefore, it is advisable to express the abundance of a constituent directly by a number,

Table 4.2. Proposed adjectives for expressing the abundance of fabric units.

Abundance	% by area
Very dominant	> 70
Dominant	51–70
Frequent	31–50
Common	16–30
Few	5–15
Very few	< 5

for example about 10% gypsum. For general purposes the adjectives indicated in Table 4.2 are proposed.

Note that the abundance of clay coatings in a thin section will usually be "very few", although 5% coatings already points to an extreme illuviation.

For most purposes in describing thin sections, a visual estimate is sufficient, often with the aid of specially designed graphs (Fig. 4.12). Several micromorphometric methods are used to determine the abundance of constituents in thin sections, ranging from the traditional point counting techniques (with point counting ocular or automatic stage) to quantitative image analysis on digital images. Discussion of this topic is beyond the scope of this manual but one should bear in mind that factors such as illumination and magnification used will influence the results. Also the experience of the analyst plays a very important role, as demonstrated by McKeague et al. (1980). Especially in the case of quantitative image analysis, several parameters such as threshold settings strongly influence the results (Marcelino et al., 2007).

Determination of abundance of constituents and fabrics is only possible when specific qualitative criteria are met (Stoops, 1977, 1978b). The units (these are the partial fabrics to be quantified) should comply with the following criteria:

1. Boundaries of the units should be sharp at the highest scale of observation applied; for instance diffuse matrix features (Section 8.2.2), such as impregnative nodules or hypocoatings cannot be quantified using areal methods.

2. Units should be homogeneous at the scale of observation with respect to their definition. For instance, sharp impregnative iron oxide nodules can be considered as a unit, notwithstanding they are complex fabrics containing both coarse material and micromass.

3. Units should be mutually exclusive. A constituent cannot be at the same time a unit and part of another, more complex unit. For instance, it is not possible to quantify at the same time quartz grains and impregnative nodules containing quartz grains. This implies that the units should belong to a same level of organization.

For pedochemical calculations only pure units (e.g., gypsum crystals, calcite grains, also fine clay coatings) can be taken into consideration. Based on their volume percentage, and knowing their density, their weight percentage can be calculated and compared with chemical data. For example, the percentage of clay in coatings calculated can be compared with the fine clay content in granulometric analyses to determine the real illuviation (e.g., Murphy and Kemp, 1984).

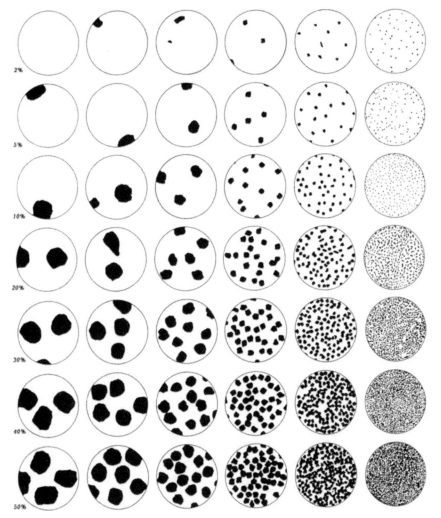

Fig. 4.12. Abundance of black objects as a percentage of visual fields with various particle sizes. Abundances greater than 50 percent are estimated from the white areas of the complementary fields. After FitzPatrick, 1984 (from Bullock et al., 1985)

4.3.6. Shape

4.3.6.1. Introduction

Shape is an important element of fabric because it may yield critical information on the origin of materials and pedogenic processes. For example, the rounding of mineral grains often points to translocation, while diffuse irregular boundaries of nodules are an indication of in situ formation. Shapes of crystals can give useful information on the environment during their formation (e.g., different crystal forms of gypsum or pyrite).

In the case of individual minerals that display clear crystallographic faces, the terminology for describing crystal forms is recommended as far as they can be recognized in thin sections (see Section 3.1.2). Such terms include for example *hexahedron, octahedron,* and *tetragonal prismatic combined with pyramids.* For more precise reporting, Miller indices can be indicated.

In most thin section descriptions, use of general terms is satisfactory. The particles can be characterized by a particular habit such as *platy, fibrous, lenticular, columnar, acicular and tabular* (Fig. 4.13). In addition, terms like *bothryoidal, reniform, dendritic, globular, rod-like, blocky, vermiform, cylindrical,* may be useful. All these terms refer to three-dimensional realities and therefore should be used with caution. The shape of the fabric units in thin sections can only be described according to two dimensions, but it is sometimes possible to infer or even deduce the three dimensional form.

For fabric units other than euhedral or subhedral crystals, the description of shape is only based on (i) equidimensionality, (ii) degree of roundness, (iii) rugosity, and (iv) boundary.

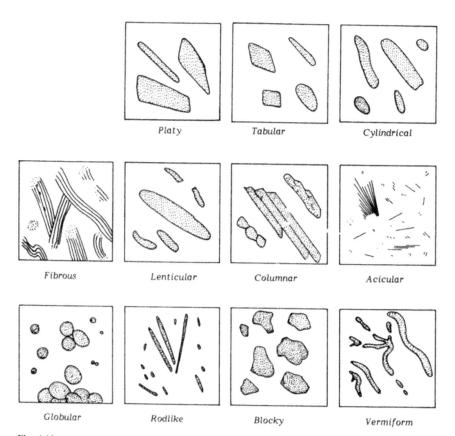

Fig. 4.13. Some shapes of features seen according to different sections. (from Bullock et al., 1985).

4.3.6.2. Equidimensionality

Equidimensionality expresses how equal are the three mutually perpendicular dimensions of a particle, by giving the relation between the sizes of the three perpendicular axes. Four classes: *oblate* (disk), *equant* (spheroid), *triaxial* (blade), and *prolate* (rod), were defined by Zingg (1935) to which Brewer (1964a) added: planar, acicular, and acicular-planar. The terms, which apply to three-dimensional forms, are defined below and illustrated in Fig. 4.14. Some modifications have been made to the class limits.

Equant (or compact): the three axes are of the same order of magnitude.

Prolate: one of the axes is considerably longer than the others.

Acicular (needle-shaped): one axis is much longer that the others.

Planar: one of the axes is considerably shorter than the others.

Acicular-planar (**lath shaped**): one dimension is much longer and the second much shorter than the third.

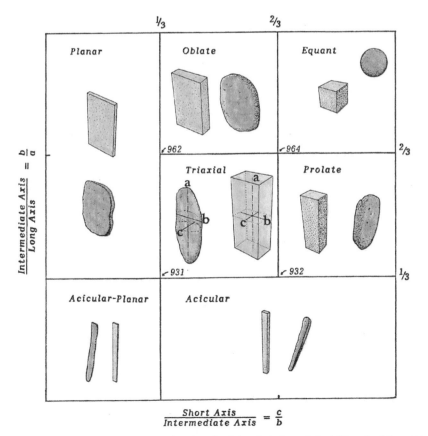

$$\frac{Short\ Axis}{Intermediate\ Axis} = \frac{c}{b}$$

Fig. 4.14. Soil feature shape classes based on the ratio of the principal axes (a, b, c). The ratio of the axes a: b: c are indicated for example as 931 (from of Bullock et al., 1985).

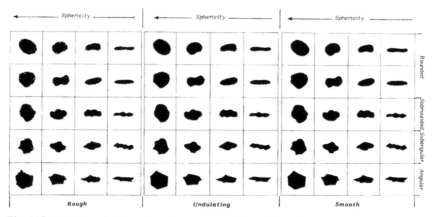

Fig. 4.15. Soil feature sphericity and roundness charts combined with roughness/smoothness grades (from of Bullock et al., 1985).

When determining the equidimensionality of an object in thin section, one must take into account the problems caused by the transition from two to three dimensions, as explained in Section 3.1.2.

4.3.6.3. Degree of Roundness and Sphericity

Roundness is defined as the relative sharpness of the particle corners. Five classes are used (Fig. 4.15), following Pettijohn (1957):

Angular: strongly-expressed faces with sharp edges and corners; secondary corners numerous and sharp.

Subangular: strongly-expressed faces with somewhat rounded edges and corners; secondary corners numerous.

Sub-rounded: poorly-expressed flat faces with corners well rounded. Secondary corners numerous.

Rounded: flat faces almost absent with all corners gently rounded; secondary corners greatly subdued and few.

Well-rounded: the entire surface consists of broad curves; secondary corners are absent.

Several visual aid charts for estimating roundness exist, for example Russell and Taylor (1937), Powers (1953), Krumbein and Sloss (1963), Hodgson (1976).

Sphericity refers to the overall form of the particle irrespective of the sharpness of the edges or corners. Accurate determinations of sphericity are tedious, and as with roundness, visual estimation charts are commonly used. Figure 4.15 combines both sphericity and roundness.

4.3.6.4. Surface Roughness and Smoothness

Roughness of the surface is an important characteristic for many fabric units, such as aggregates, voids, excrements, and nodules, as it can

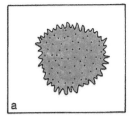 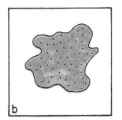 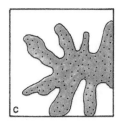

Fig. 4.16. Examples of boundary roughness of soil features: (a) serrate, (b) mammilate, (c) digitate. (after Bullock et al. 1985).

yield information on the genesis or alteration of these units. Two levels of roughness and smoothness can be considered, one at a low magnification, describing the overall irregularities of the surface, and one at a higher magnification, at the level of basic constituents. At the low magnification, the depth to which the surfaces are broken up by indentations, projections, or protuberances and their shape is considered. The following types are distinguished: **regular, mamillated, crenulate, palmate, digitate,** and **serrate** (Fig. 4.16).

At the higher magnification, the surface may seem

Rough: the surface has indentations deeper than they are wide.

Undulating: the surface is dominated by broad shallow undulations.

Smooth: there are few or no irregularities on the surface.

The degree of roughness or smoothness depends on magnification. A given fabric unit may appear to be smooth at a low magnification, but rough at a higher one. It is therefore important to state the magnification used.

4.3.6.5. Boundary

Two properties, sharpness and contrast, are important in relation to the boundary between the particle or feature and the adjacent material. Sharpness relates to the transition between the feature and the matrix and other features; contrast refers to the degree to which the feature is clearly differentiable from other features and the matrix. It is evident that the characteristics of a boundary are a function of the magnification. Boundaries that are rather gradual at high magnification may appear sharp at a lower one. Therefore it is necessary to specify the magnification used.

Three degrees of sharpness (at a specified magnification both in PPL and XPL) can be distinguished.

Sharp: boundaries between colors and/or particle size distributions show no transition.

Clear: color and/or particle size transition < 60 µm wide.

Diffuse: color and/or particle size transition > 60 µm wide.

Three degrees of contrast (in PPL and XPL) are distinguished by Bullock et al. (1985):

Prominent: the individual is conspicuous and stands out from other individuals in terms of color, particle size-distribution, interference colors, or other morphological properties.

Distinct: although not striking, the individual can be seen clearly. It has some morphological properties in common with other individuals. In terms of color, the individual usually has the same hue as the one with which it is compared but differs in chroma.

Faint: the individual is only evident on close examination. In terms of color, the individual and the feature with which it is compared have the same hue and differ only slightly in chroma. Similarly, there will be only a small difference in particle size distribution between the two features.

In some cases, pedofeatures for instance, a more extensive contrast scale may be appropriate. This can be achieved by adding the prefix very to the terms prominent and faint.

Contrast depends on magnification and type of light used. In many cases contrast is better expressed at low magnifications than at high magnification. Inserting the swing out (substage) condenser generally decreases contrast caused by color.

4.3.7. Color

Color described in transmitted light is a function of several parameters: (i) thickness of the section, (ii) temperature (color) of the light used (yellow-white, white, or blue-white), magnification, and type of condenser used.

Colors of transparent bodies are always described in PPL. Attention should be given to the presence of pleochroism, for example in minerals or clay coatings.

In the case of cryptocrystalline or amorphous materials, such as a clayey groundmass, clay coatings, or Fe-oxide nodules, the thickness of the thin section and the magnification strongly influence the observed color (Plate 3.1c and d). The reddish color of the clayey groundmass observed in thin sections 30 μm thick might become gradually more yellowish in thinner ones, and darker, up to opaque, in thicker ones. At low magnification, nodules may appear black and opaque, but brownish or reddish at high magnification with the swing out condenser inserted. It is therefore useful for these materials to describe the color, also in OIL or with a dark-field condenser (Plate 3.2c and d).

The description of color as a diagnostic characteristic is especially useful in the case of cryptocrystalline and amorphous materials.

Care should be taken when observations are made on a PC screen, because then colors depend also on the settings. Any digital imaging should be calibrated against know reference colors (e.g., Munsell). Therefore, colors are better described through the eyepiece.

4.4. VARIABILITY WITHIN PARTIAL FABRICS

Variability deals with the variation existing among individual fabric units of a same type, such as among quartz grains, nodules, or peds. Bullock et al. (1985) proposed the following scale of variability:

Low: weak differences in one characteristic between individuals of a same partial fabric.

Medium: moderate differences in one characteristic or weak differences in several characteristics between individuals of a same partial fabric.

High: strong differences in one characteristic, moderate differences in several characteristics, or weak differences in many characteristics between individuals of a same partial fabric.

4.5. CONCEPTS USED

Soil is an uttermost complicated material. Its fabric cannot be studied in a single approach. For this reason a distinction is made by Bullock. et al. (1985) between groundmass and pedofeatures, as was done already by Brewer (1964a) who distinguished s-matrix and pedological features. The last term was rejected, because it means in fact "features of soil science", and is therefore moreover difficult to translate.

Pedofeatures are discrete fabric units present in soil materials recognizable from an adjacent material by a difference in concentration in one or more components (e.g., a granulometric fraction, organic matter, crystals, chemical components) or by a difference in internal fabric (Bullock et al., 1985).

Groundmass is a general term used for the coarse and fine material and associated packing void, which forms the base material of the soil in thin section, other than that in pedofeatures (Bullock et al., 1985, modified).

For a more thorough analysis of the concepts, refer to Section 7.1.

5. Voids, Aggregates and Microstructure

5.1. INTRODUCTION

In field descriptions (Schoenberger et al., 1998) soil structure is considered as the natural arrangement of soil particles in aggregates due to pedogenic processes. Soils without aggregates are considered, therefore, as structureless. This is the case when the soil has a coherent aspect, and then it is described as massive, or when the material is not coherent, such as sandy materials. The presence of aggregates in soils automatically involves the presence of voids, but the opposite is not true: voids may be abundant in a soil without peds being formed. In field systems, voids are not taken into consideration when discussing soil structure; in micromorphology, on the contrary, porosity is part of microstructure (Bullock et al., 1985; Stoops, 2003). This means for example that a coherent material (described as massive in the field) can be described for instance as a channel microstructure in soil micromorphology.

To maintain links with field descriptions, Bullock et al. (1985) proposed the term structure or microstructure, for that part of fabric related to aggregation and/or porosity. In agreement with the definition of fabric used in Section 4.2 of this book, the following definition of structure, which is applicable to both aggregated and non-aggregated material, has been adopted:

"*Soil microstructure* is concerned with the size, shape and arrangement of primary particles and voids in both aggregated and non-aggregated material and the size, shape, and arrangement of any aggregates present" (Bullock et al., 1985, modified).

Soil structure (or microstructure) is thus that part of soil fabric dealing only with the relation between the solid and the voids. It is not restricted to pedality, but also implies apedal structures. "Massive" in this sense does not refer to a material without pedality, but to a

Guidelines for Analysis and Description of Soil and Regolith Thin Sections, Second Edition. Georges Stoops.
© 2021 Soil Science Society of America, Inc. Published 2021 by John Wiley & Sons, Inc.
doi:10.2136/guidelinesforanalysis2

material without porosity. The concept fits rather well with that of soil structure as used in the field. As defined above, structure does not include the internal fabric or composition of the solid phase. However, the internal fabric of aggregates may be an important factor for interpretation, and should be described as part of the groundmass. In some Russian publications internal fabric is one of the criteria for differentiating aggregate types (Gerasimova et al., 1996; Gerasimova, M. & Lebedeva-Verba, 2018).

Bullock et al. (1985) considered microstructure as all aspects of soil structure revealed when a soil material is examined at a magnification of five times or more; as such, many macroscopic structural features are also included.

The description of the microstructure of a pedal material comprises both the description of the aggregates and the inter- and intrapedal voids; in an apedal material, the morphology and pattern of the voids describe the microstructure. Following the approach of Beckmann and Geyger (1967) a combination of porosity and aggregation finally leads to the different types of microstructures (Section 5.4).

BACKGROUND - In his book Micropedology, Kubiëna (1938) used structure, combined with other features, at a high level to distinguish seven main types of fabric. Structure as such was not discussed in his work. In Brewer (1964a), the discussion of microstructure, in terms of field usage, was restricted to pedality as described in soil survey manuals. His important contribution to microstructure descriptions was in the morphological classification of void (pore) types. Combining aggregate- and pore types, Beckmann and Geyger (1967) were the first to define a series of microstructures. Although all void and aggregate types were not considered, and grain structures were completely omitted, the concept was an excellent basis that served as a model for Bullock et al. (1985). Brewer and Sleeman (1988) tried to address with the problem of linking the discontinuous grain microstructures (corresponding to nonporphyric c/f related distributions) with more continuous microstructures, introducing a set of newly coined terms.

5.2. VOIDS

5.2.1. Introduction

Voids, as understood here, are spaces not occupied by solid soil material. They determine the porosity in the soil, a fundamental property from a global environmental point of view, as it determines the water and air cycles, and as such soil and life on Earth. Its relation to soil fauna and flora is reciprocal: porosity influences the development of organisms in and on the soil, which in turn create porosity (Morrás, 2015). Most pedogenic processes are related to the presence of voids.

Porosity in the soil forms a continuum, except in a few special cases, (e.g., vesicles). Because this is practically impossible to describe, Brewer (1964a) classified voids according to their morphology, considering them as individuals. Beckmann and Geyger (1967) produced a more simple classification with only two classes (fissures and cavities), each of which were subdivided using adjectives to denote particular characteristics; no packing voids or channels were considered.

To emphasize that in thin sections one is dealing with two-dimensional sections through three-dimensional voids, Moran et al. (1988) coined the term *poroids* for the two-dimensional sections. When necessary for clarity, this term will be used in this text and the key. Using image analysis techniques on multiple parallel thin sections or micro CT scanning, a three dimensional image of the porosity can be obtained.

Since voids between finer constituents (e.g., clay particles) in thin section are beyond the resolution of the optical microscope, they are not considered in this description scheme.

5.2.2. Types of Voids (Fig. 5.1)

Packing voids: voids resulting from the loose packing of soil components, the faces of which do not accommodate. They are equant to elongate, much interconnected, and form the so called textural porosity of the soil (see Section 5.4.1). They can be:

Simple packing voids: between basic components (e.g., sand grains) (Plate 4.4a and b, Plate 4.5e);

Compound packing voids: between nonaccommodating aggregates, (e.g., between granules or crumbs) (interaggregate voids) (Plate 5.2, Plate 8.18a through c);

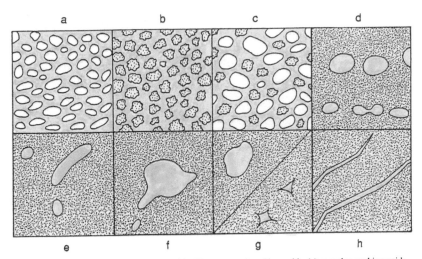

Fig. 5.1. Types of voids: (**a**) simple packing voids, (**b**) compound packing voids, (**c**) complex packing voids, (**d**) vesicles, (**e**) channels, (**f**) chamber, (**g**) regular and star-shaped vughs, and (**f**) planes.

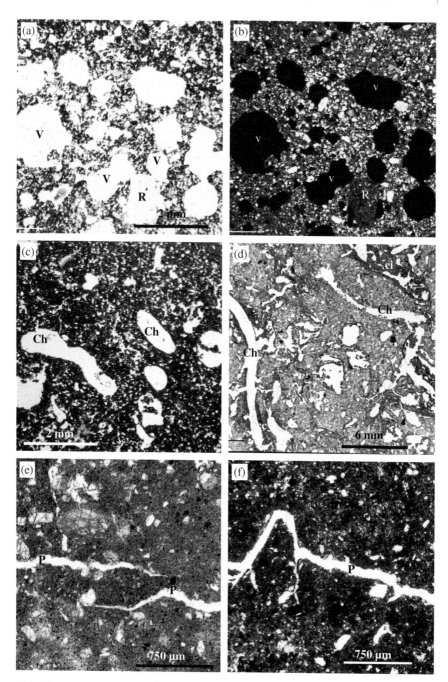

Plate 5.1. *Types of voids* (**a**) Vesicules in crust (V: vesicle; R: rock fragment) (PPL), (**b**) same; notice crystallitic b-fabric of micromass (V: vesicle; R: rock fragment) (XPL), (**c**) channels (ch) forming a channel microstructure superposed to weakly developed granular microstructure (PPL), (**d**) channel (ch) microstructure with coatings (cc) and infillings (ci) of illuvial clay (PPL), (**e**) subparallel accommodating planes (p) ending in sharp points (PPL), (**f**) accommodating plane (p) (PPL).

Complex packing voids: between basic components and small aggregates (e.g., in materials with an enaulic c/f related distribution) (Plate 4.5a, b and c).

Vesicles: relatively large voids whose walls consist of smooth, simple curves; equant, prolate, or oblate. In vertical sections, vesicles usually have a (sub)horizontal referred distribution. These voids have been attributed to the incorporation of air bubbles in near surface horizons (e.g., in vesicular crusts) (Plate 5.1a and b), but were observed by the author also in the puddled layer of paddy soils.

Channels: tubular smooth voids with a cylindrical or arched cross-section which are uniform over much of the length; they are mainly root channels or biogalleries. In thin sections they appear as rounded or short prolate poroids with rounded edges, and a circular, ellipsoid or arched cross-section (see also Fig. 3.2.) (Plate 3.2c and d, Plate 3.4a and b, Plate 5.1c and d, Plate 5.5b).

Branching patterns of channels, as proposed by Brewer (1964a) are not described here, as they refer to a three-dimensional reality not visible in thin sections.

Chambers: more or less equidimensional smooth walled pores interconnected by channels; in general they are not easily recognized as such in thin sections because only exceptionally the interconnection will be situated in the plane of the section.

Planes: planar (according to the ratio of principal axes, one being much smaller than the two other), flat voids, accommodating or not, smooth or rough; they are the result of shrinkage or slipping (Plate 5.1e and f, Plate 5.3). In thin sections, they appear generally as elongated poroids terminating in at least one sharp edge (Plate 5.1e), as opposite to channels. The presence and size of planes is influenced by the technique used to extract the water from the sample before impregnation: air- and oven drying generally results in more and wider planes.

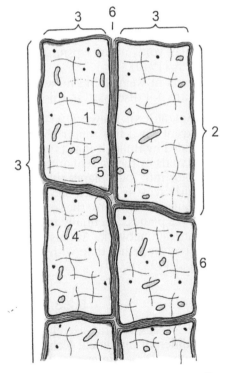

Fig. 5.2. Hierarchy in microstructure: (1) primary, weakly separated angular blocky peds, (2) highly-developed secondary accommodating angular blocky peds, (3) highly-developed tertiary prisms, (4) intra- and probably transpedal channel, (5) intrapedal vugh, (6) interpedal coating, and (7) intrapedal nodule.

According to their overall shape, three main subtypes can be distinguished:

- *Straight planes*: those that maintain more or less the same direction over their full length;

- *Zigzag planes*: those that change direction more than once throughout their length;

- *Curved planes*: curved, sometimes even circular (as in onion skin microstructures or around hard grains or nodules) (Photos Plate 5.5c).

The degree of accommodation, of planar voids is a measure of the degree to which opposite faces exhibit complimentary shapes (see Section 5.2.7) (Plate 5.1e and f).

Vughs: more or less equidimensional, irregular voids, smooth or rough, normally not interconnected to voids of comparable size; they result from the welding of aggregates, disruption of the microstructure or dissolution of components (Plate 5.1d, Plate 5.5a). Polygonal poroids (mostly triangular or quadratic) with convex walls, resulting from the welding of rounded aggregates are called star-shaped vughs.

Moldic voids: voids pseudomorph after euhedral or suhederal minerals, resulting from the congruent dissolution of minerals (e.g., lenticular voids after gypsum). The term was proposed by Choquette and Pray (1970).

Voids in weathering minerals, such as contact voids (see Section 6.2.2) are not discussed in this chapter.

> BACKGROUND- Brewer (1964a) defined void types in a similar way, except for a few minor changes. Complex packing voids were added by Bullock et al. (1985), and star-shaped vughs and moldic voids by the present author. Brewer (1964a) subdivided planar voids into joint planes that transverse the soil material in some fairly regular pattern such as parallel or subparallel sets, skew planes that traverse the soil material in an irregular manner, and craze planes, essentially irregular planar voids that occur as an intricate network. As this subdivision in fact deals with microstructure rather than with void types, it was omitted by Bullock et al. (1985).

5.2.3. Size of Poroids

In soil physics, several size classifications for pores exist that are related to different analytical techniques. As they refer to the more or less continuous three-dimensional reality, they are not suitable for micromorphological studies. Therefore, it is suggested the size of the voids be expressed directly in micrometers or millimeters.

The size of the packing voids is directly related to that of the composing grains and/or aggregates. For channels and vesicles, the diameter should be recorded. For chambers and vughs, measurements of length and width are necessary. For planes, the thickness (minimum and maximum) must be reported. One should be aware that the width of planes can be influenced by the drying procedure, and by the inclination with regard to the section (see Section 3.1.3). The estimation of the pore-size distribution is an important parameter. Today image analysis is used more and more to determine poroid shape, size and distribution.

5.2.4. Abundance of Voids

The following abundance estimates are useful for characterizing voids: (i) the total void space as a proportion of the thin section in area, and (ii) the proportions of the different void types as a proportion of the total void space.

5.2.5. Roughness and Smoothness of Void Walls

The roughness and smoothness of void walls is an important characteristic, both from a genetic and a practical point of view, as explained in Section 5.3.6. The terminology explained in Section 4.3.6.4 should be used. Two levels can be distinguished: (i) at the scale of the groundmass (e.g., mammilated vughs) and (ii) at the scale of the coarse and fine material in the groundmass, such as protrusion of coarse grains (e.g., in the case of shrinkage planes) or smoothened walls (e.g., in the case of channels and slickensides).

> BACKGROUND - Brewer (1964a) using the second level, distinguished *ortho-* (rough) and *meta-* (smooth) voids (e.g., metachannels, orthovughs). These terms are not recommended, especially because the same author uses the terms ortho-, meta-, and para- also to indicate the in situ or inherited origin of some features.

5.2.6. Arrangement of Voids

When not randomly distributed and/or oriented, the pattern should be described using the concepts and terms explained in Section 4.3.2. In pedal soils, three related distribution patterns can be recognized (Fig. 5.2, Plate 5.3a and b): (i) between aggregates (interpedal), (ii) within aggregates (intrapedal) and (iii) across aggregates (transpedal). It is important to recognize these three situations and to describe them separately.

5.2.7. Accommodation of Voids

The accommodation of voids is a measure of the degree to which opposite faces exhibit complimentary shapes, in other words shapes that can be fitted or meshed together by an imaginary movement or translation that brings the opposing faces together (see also Section 5.3.5). However, it should be kept in mind that movements do not necessarily have to take place in the plane of the section. Bullock et al. (1985), following Brewer (1964a), defined three degrees of accommodation (Fig. 5.3):

Accommodated: all faces are accommodated by the faces of adjoining aggregates (Plate 5.1e and f).

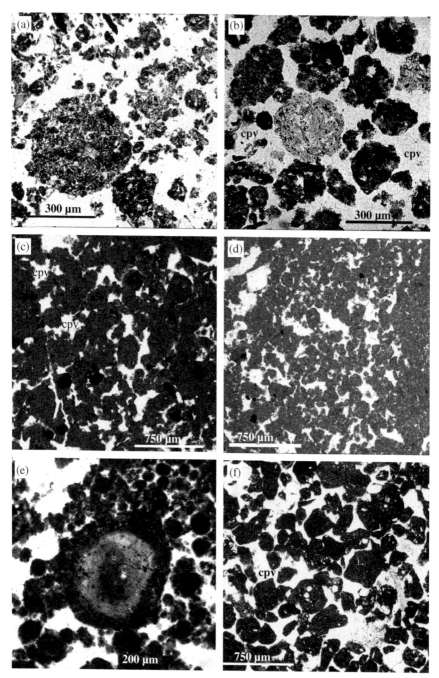

Plate 5.2. *Spheroidal peds.* (a) crumb microstructure characterized by spheroidal porous aggregates (crumbs) and compound packing voids (PPL), (b) Crumb microstructure with porous spheroidal aggregates and compound packing voids (cpv); the aggregate in the center is an excrement (e), composed of fresh cell fragments (PPL), (c) moderately separated granular microstructure with compound packing voids (cpv) (PPL) (d) moderately separated (center) and weakly separated (right) granular to vughy microstructure with star shaped vughs (PPL), (e) concentric granular ped in volcanic ash soil (PPL) (f) angular micropeds and compound packing voids (cpv), often described as a highly separated granular microstructure, (PPL).

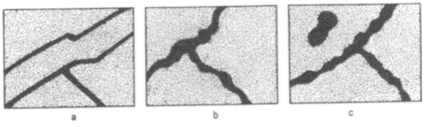

Fig. 5.3. Degree of accommodation: (a) accommodated, (b) partially accommodated, and (c) unaccommodated.

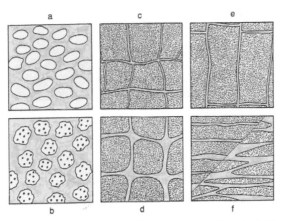

Fig. 5.4. Morphological types of peds: (a) granules, (b) porous crumbs, (c) angular blocky peds, (d) subangular blocky peds, (e) prisms, (f) plates and lenticular plates.

Partially accommodated: some faces are accommodated by faces of adjoining aggregates.

Unaccommodated: virtually none of the faces are accommodated by the faces of adjoining aggregates; adjoining aggregates touch only at points.

5.2.8. Note

In fabric analysis of soils, the term void is sometimes used in two different ways: (i) as a pore, in the physical sense, which can be filled with liquid and/or gas, as used in the description of microstructure, and (ii) as a fabric element that has once been a pore in the groundmass, but which later became totally or partially filled by pedofeatures such as coatings or infillings. Thus a channel in this sense can be partly filled by loosely packed aggregates (Plate 8.19a and b). In this example the resulting physical pore space corresponds to compound packing voids, but the micromorphologist will still consider it morphologically as a channel. This is important, as it will have an impact on the total permeability, as stated by Vepraskas et al. (1991). Some researchers therefore put a limit to the amount of solid material allowed in a given void type (e.g., Vogel, 1994). Therefore, in the description of voids, one has to distinguish between functional voids,

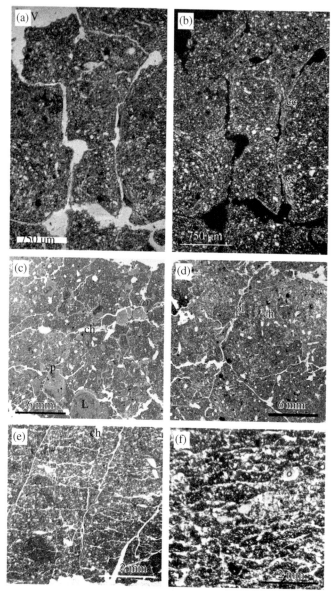

Plate 5.3. *Blocky peds and plates* (**a**) highly separated subangular blocky microstructure with mainly accommodating planes (v), interpedal clay coatings (cc) and intrapedal diffuse Mn-hydroxide nodules (n) (PPL); (**b**) same; the interpedal clay coatings (cc) are more evident than in PPL (XPL); (**c**) angular blocky microstructure with accommodating planes, intrapedal channels (ch) and many gray fragments of limestone (L); notice shrinkage planes (p) around the chalk fragments (PPL); (**d**) angular blocky microstructure with planar voids (pl) and some channels (ch).(PPL); (**e**) highly separated angular blocky microstructure with tendency to platy microstructure; accommodating planes; few channels (ch) (PPL); (**f**) lenticular platy microstructure in subsurface horizon of temporarily frozen soil (PPL)

corresponding to the empty spaces, and genetic voids, determining the shape of the former and present voids.

5.3 AGGREGATION

5.3.1. Introduction

In field studies, a distinction has been made between natural, relatively permanent aggregates (peds), separated from each other by voids or natural surfaces of weakness, and less permanent, mainly artificial aggregates (fragments and clods) resulting from cultivation or frost action in surface horizons (Soil Survey Staff, 1993). Because the distinction between peds and clods or fragments has a genetic rather than a morphological base, it is difficult to distinguish between both types of aggregates in thin sections. Moreover such a distinction would represent an interpretation rather than an objective description. For this reason all aggregates are described as peds in this text, contrary to what was done by Bullock et al. (1985).

5.3.2. Peds

The Soil Survey Staff (1993) recognized four main types of peds: granules, plates, blocks, and prisms. All can be recognized at a microscopic level, although prisms will generally not be visible because of their large size compared to the common size of thin sections. The following four main ped types (Fig. 5.4) are distinguished:

Spheroidal peds: particles are arranged into more or less equant peds that are bounded by rounded faces and, as a consequence, are not accommodated to the adjoining peds. There are two types: *crumbs* (Plate 5.2a and b) and *granules* (Plate 5.2c through f, Plate 5.4a and b), the former being porous and the latter nonporous at the scale of the optical microscope. Crumbs and granules, although considered as spheroidal peds, do not necessary have a perfect spherical shape. In practice, loose subangular micropeds are often described as granules (Plate 5.2f). Granules are often related to excrements (see also Section 8. 8).

Blocky peds (Plate 5.3): particles are arranged into more or less equant peds that are angular if bounded by flat surfaces and subangular if bounded by flat and rounded surfaces. Faces of the *angular blocky peds* are more or less mirror images of the faces of the surrounding peds, separated from them by accommodation planar voids, except where the soil has been disturbed by cultivation or where the peds have accumulated in voids as the result of natural fracturing and/or gravitational transfer. Surfaces of *subangular blocky peds* can also be molds of each other, but due to high biological activity or cultivation they may be loosely packed and only touch each other in some parts.

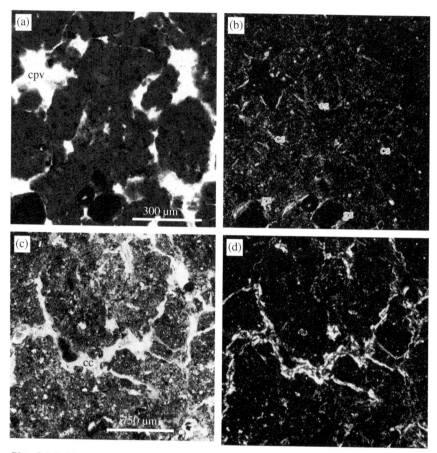

Plate 5.4. *Pedality and degree of separation.* (**a**) medium separated granular microstructure; the micromass of the aggregates has a dotted limpidity; compound packing voids (cpv). (PPL); (**b**) same, notice the circular striated b-fabric (cs) in the center and the granostriated b-fabric around opaque grains (gs), superposed to a weakly developed stipple speckled b-fabric (XPL); (**c**) strongly developed, but weakly separated angular blocky microstructure; the interpedal voids contain dense complete infillings or intercalations of yellow illuvial clay (ci) (PPL); (**d**) same, notice interference colors of clay infilling and the extinction lines (XPL).

Plates: particles are arranged into generally elongate (sub)horizontal peds. There are three main sub-types:

- *More or less straight plates* (Plate 5.3e).
- *Wavy plates*, as present in some frost affected soils.
- *Lenticular plates,* which are thick in the middle and thin toward their edges, characteristic for frost affected soils. (Plate 5.3f, Plate 8.22a and b).

Prisms: particles are arranged into more or less vertically elongate peds bounded by surfaces that are usually flat, but which can also be rather rough. They are subdivided into prisms with caps (columns) and without caps (prisms). Prisms and columns are generally too large to fit in a thin section.

A hierarchy of microstructures may be observed in peds (Jongerius, 1957). For instance, a prism may be composed of large angular blocky peds, which in turn separate into smaller blocky peds. The latter may display a distinct apedal internal microstructure, e.g., a channel microstructure. Some authors consider the largest units observed as "primary peds" (e.g., Bullock and Murphy, 1976), while others assign the smallest ones to that category (Fig. 5.2) (e.g., Brewer, 1964a; Bullock et al., 1985). In this book, the first approach is followed (see also Section 5.4.1), because the limited size of thin sections does not always allows to observe the largest peds.

5.3.3. Degree of Ped Separation and of Pedality

In the system used by the USDA, peds are separated by zones of weakness that may be easily detected in the field, but which are not obvious in thin sections, where the ease of dislocation of a ped cannot be observed (Soil Survey Staff, 1993). This same manual defines grade of structure as the degree of aggregation, and expresses the differences between cohesion within the aggregates and adhesion between the aggregates. Several micromorphologists (FitzPatrick, 1984; Bullock et al., 1985) proposed a micromorphological definition of the grade of pedality based on the only characteristic visible in thin sections, namely the degree to which the aggregates are surrounded by voids. Systematic application of this criterion, in comparison with field descriptions, showed serious discrepancies in many cases. Soils with a high degree of pedality in the field were not evaluated as such in thin sections, as pedality is not only a factor of separation. R. Langohr (personal communication, 1996), therefore, proposed that the additional separate criteria of *degree of completeness and distinctness* be distinguished in the field. The degree of completeness is covered by the criteria proposed by Bullock et al. (1985) for pedality. In this text the term *degree of ped separation* will be used for this aspect. Three degrees of separation can be distinguished (Fig. 5.5):

Highly separated: the soil material is divided into a number of units, each of which is entirely surrounded by a void (Plate 5.2a, b and f, Plate 5.3e and f, Plate 5.5b).

Moderately separated: the soil material is divided into a number of units that are surrounded on at least two-thirds of their periphery by

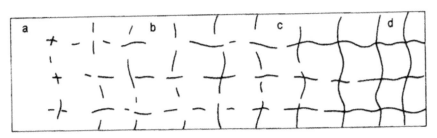

Fig. 5.5. Degree of separation: evolution of apedal material (a) over weakly separated (b) and moderately separated (c) to highly separated (d).

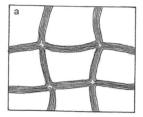 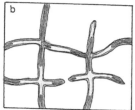 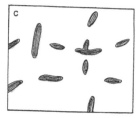

Fig. 5.6. Degree of pedality in the case of planar voids partially filled with clay coatings: (**a**) well developed pedality, no separation; (**b**) moderately developed pedality, weakly separated; and (**c**) weakly developed pedality, no separation.

planar voids and/or are attached to other peds by links less than one third the size of the mean diameter of the ped (Plate 5.3a through d).

Weakly separated: the soil material is divided into a number of units separable on the basis of being partially surrounded by planar voids (one-third to two-thirds of the periphery of the proposed unit) and/or by links between units which are one-third to two-thirds the maximum width of the unit (Plate 5.2c and d, Plate 5.4c and d).

The degree to which peds are separated by open voids (degree of ped separation) is only one of the possible criteria used to distinguish pedality. Pedality may also be identified by the presence of surfaces of weakness, such as (clay) coatings that fill large voids, or the presence of slickensides recognizable as a preferential orientation of the clay in the groundmass (porostriated b-fabrics, see Section 7.2.4.2). It is therefore suggested that the subdivision proposed by Bullock et al. (1985) be used after adding the following criteria (Fig. 5.6):

Strongly developed pedality: the soil material is divided into a number of units, each of which entirely surrounded by a void or a feature pointing to a surface of weakness (Plate 5.4).

Moderately-developed pedality: the soil material is divided into a number of units which are surrounded on at least two-thirds of their periphery by voids or features pointing to a surface of weakness, and/or are connected to other peds by links less than one-third the size of the main diameter of the ped.

Weakly-developed pedality: the soil material is divided into a number of units separable on the basis of being partially surrounded by voids or features pointing to a surface of weakness (one third to two thirds of the periphery of the proposed unit), and/or by links between units that are one-third to two-thirds the maximum width of the unit.

In the case of moderately- and weakly-developed peds, the voids and/or links are inferred to be the points at which aggregates will separate if pressure is applied.

5.3.4. Size of Peds

Bullock et al. (1985) suggested that the most commonly used field size classes for aggregates should be adopted to have continuity between field and microscopic descriptions. Experience has shown, however,

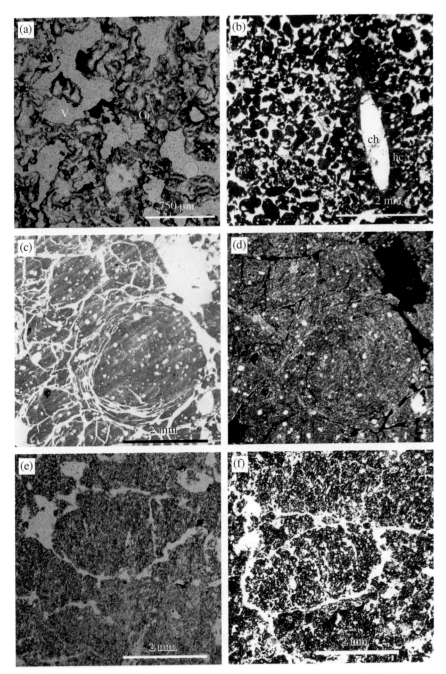

Plate 5.5. *Microstructures.* (a) spongy microstructure in gibbsitic horizon of bauxite profile (v: voids; Gi: gibbsite) (PPL), (b) highly separated granular microstructure with channel (ch), surrounded by fabric (compaction) hypo-coating (hc) (PPL), (c) spheroidal (onion skin) microstructure (PPL), (d) same, note generalized strong striated b-fabric (st) (XPL), (e) vermicular microstructure, consisting of an accumulation of large organo-mineral excrements (PPL), (f) vermicular microstructure, consisting of porous excrements; notice crescent fabric in central aggregate (PPL).

that this approach is not practical because the finest granules in this scale still belong to a very coarse type in microscopic investigations, limiting further subdivisions. It is therefore proposed that aggregate sizes be indicated in units of micrometers, millimeters, or centimeters. For fine granules, terms such as medium or coarse sand size can be used.

5.3.5. Accommodation

Accommodation of peds to each other is a measure of the degree to which opposite faces exhibit complimentary shapes (see Section 5.2.7).

For more detailed descriptions, the degree of accommodation can be expressed as a percentage of matching surfaces. Spheroidal peds are by definition unaccommodated, whereas prismatic, platy, and in situ angular blocky peds normally have a high degree of accommodation. Subangular blocky peds may be at least partially accommodating, when in situ. Blocky peds occurring as infillings (see Section 8.4.) or in cultivated or strongly bioturbated layers are no longer accommodated.

5.3.6. Surface Roughness

A description of this characteristic is important both from a genetic and a practical point of view. In the case of spheroidal aggregates (e.g., excrements) surface roughness may yield some information about genesis and evolution. In the case of blocky peds, prisms, and plates, roughness may help to predict water and air movement, since rough surfaces do not allow complete closing of planes on swelling, whereas smooth accommodating surfaces more effectively seal and prevent water movement by allowing better contact on swelling. The degree of surface roughness and smoothness of aggregates should be described using the three categories given in Section 4.3.6.4.

5.3.7. Internal Fabric

Although the internal fabric of aggregates is not part of their definition, its description can be very important in some specific cases, using concepts and terms proposed in Chapter 4. Examples are the crescent or bowlike fabric in some crumbs (Plate 5.5f), pointing to a zoogenetic origin (generally earthworm excrements), or concentric granules in young volcanic ash soils, inherited from the parent material (Plate 5.2e) (Schumacher and Schmincke, 1991 and 1995).

5.3.8. Ped Arrangement Patterns

The general terminology explained in Section 4.3.2 is applicable to the description of the arrangement of aggregates. The most common basic distribution patterns are random, clustered, and banded. Tubular is an additional pattern applicable particularly to peds in which the individuals are grouped within channels (see also infillings, Section 8.4.).

Referred distribution and orientation patterns (with respect to the soil surface) are especially important for the description of platy aggregates.

5.4. TYPES OF MICROSTRUCTURES

5.4.1. Introduction

In reality, an infinite number of possible microstructures can be observed, and only the main types have been named and defined. Most soil materials have more than one type of microstructure at different level of observation, and a hierarchy of microstructures can be recognized (see also Section 5.3.2, Fig. 5.2). For example, some prisms will separate into an angular blocky microstructure, where each angular block will possess an intrapedal microstructure dominated by channels. Moreover, it frequently happens that two or more microstructures are juxtaposed at the same level (e.g., channel and vugh microstructure) (Plate 5.1d). Sandy materials occupy a special place with regard to microstructures, as these are determined by the related arrangement of the basic units: sand grains, fine material, and associated voids. The resulting inherent porosity is considered as textural porosity, as opposed to structural porosity (e.g., planes, channels, vesicles). This level is related to the c/f related distribution patterns and is called basic microstructure. Another subdivision of voids, has been defined for image analysis by Ringrose-Voase (1991) who distinguished structural pore space (packing pores and fissures) and non-structural porosity (the other voids).

BACKGROUND- Several attempts have been made to overcome this by devising a uniform classification of microfabrics (e.g., Brewer and Sleeman, 1988, Jim, 1988a). The drawback of such schemes is that many new-coined terms have to be added, and especially that direct (semantic) links with the field are lost. The author hopes that the introduction of subtypes for the enaulic c/f related distribution pattern may partially solve this problem.

5.4.2. Main Types of Microstructure

The following list gives the most common microstructures and their main characteristics. The list in no way covers all possibilities.

(i) **Massive microstructure**: no separated peds; few, if any, visible voids.

(ii) **Basic microstructures**: correspond to the c/f related distribution patterns in sandy soils, with exclusion of the porphyric one. As mentioned above they display an inherent textural porosity. In the case of microstructure, it is not important whether the fine material is part of the groundmass, or a pedofeature, nor whether the coarse material is lithogenic or pedogenic (e.g., lenticular gypsum crystals). For instance, in the case of pellicular grain or bridged grain microstructures, the fine material is often formed by illuvial clay or colloid coatings or bridges.

Single grain microstructure: almost exclusively sand-sized grains; little or no fine material in intergranular spaces, corresponding to a coarse monic c/f related distribution pattern (Plate 4.4a and b, Plate 4.5e); grains seem floating, without contact (see also Section 3.1.2).

Bridged grain microstructure: almost entirely sand-sized grains bridged by fine material, usually clay or organic matter, corresponding to a gefuric c/f related distribution patterns (Plate 4.4c and d).

Pellicular grain microstructure: almost entirely sand-sized grains with most grains coated by fine material, corresponding to chitonic c/f related distribution pattern (Plate 4.4e and f).

Intergrain micro-aggregate microstructure: almost entirely sand-sized grains with micro-aggregates of fine material in between, corresponding to one of the enaulic c/f related distribution patterns (Plate 4.5a through c)). As mentioned already in Section 4.3.2.4 a gradual transition exists from intergrain micro-aggregate microstructure to a granular one (Stoops, 2008) (Fig. 5.7). The gradual transition from intergrain (enaulic) over microgranular and compact microgranular to vughy (first with star shaped vughs, finally with vughs with a polyhedral shape) has been well documented by Bertran et al. (1991). The difference between microstructures with a textural porosity and those with a structural porosity creates an artificial break in the description of microstructures.

(iii) Pedal microstructures

Granular microstructure: granules are separated by compound packing voids and do not accommodate each other; they contain few

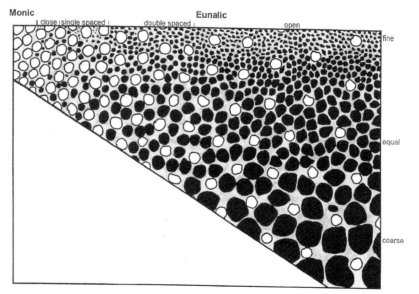

Fig. 5.7. Continuity between monic and enaulic c/f related distribution patterns and granular microstructures (From Stoops, 2008).

or no voids or recognizable smaller units (Plate 5.2c through f, Plate 5.4a and b, Plate 5.5b, Plate 6.14f).

Crumb microstructure: more or less rounded, often rugose porous aggregates not accommodating each other; the interior of the aggregates can be composed of small granules more or less welded together (Plate 5.2a and b).

Angular blocky microstructure: aggregates have angular edges, few voids and are separated by an intricate system of planar voids; faces of aggregates normally accommodate each other (Plate 5.3 a through e).

Subangular blocky microstructure: aggregates are separated by short planar voids on all or most sides; aggregate faces largely accommodate each other.

Platy microstructure: stacks of aggregates generally horizontally elongated and separated by planar voids (Plate 5.3e).

Lenticular microstructure: stacks of elongated lenticular aggregates separated by (sub)horizontal planar voids, pointing to frost action (Plate 5.3f, Plate 8.22 a and b).

(iv) Apedal and Intrapedal Microstructures

Vughy microstructure: mainly non-interconnected vughs and occasional channels and chambers.

Spongy microstructure: porous material with mainly highly interconnected vughs (Plate 5.5a).

Channel microstructure: channels as dominant voids (Plate 5.1c and d).

Chamber microstructure: chambers as dominant voids.

Vesicular microstructure: vesicles as dominant voids, generally showing a (sub)horizontal distribution (Plate 5.1a and b)

(v) Complex microstructure: mixture of two or more microstructure types (Plate 5.5b); a combination of terms can be used to name the microstructure of the whole thin section.

In addition, some specific microstructures have been described in literature:

Spheroidal microstructure (Kapur et al., 1997) called also onion-skin microstructure (FitzPatrick, 1993) consists of series of discontinuous concentric curved planar voids, mainly observed in Mediterranean soils (Plate 5.5c and d).

Vermicular microstructure (FitzPatrick 1993) consists of an intertwining mass of earthworm vermiforms. It appears as a complex system of dense complete infillings lined by continuous or discontinuous thin planar voids with irregular, circular or elliptic shapes (Plate 5.5e and f).

As mentioned in the beginning of this section, and in Section 5.3.2, a hierarchy of microstructures commonly occurs.

6. Mineral and Organic Constituents

6.1. INTRODUCTION

The simplest fabric units of soil, such as sand grains, clay particles, plant fragments, are referred to as basic components. They constitute the matrix of the soil, referred to as the groundmass as well as the matrix within pedofeatures. For practical reasons it is necessary to distinguish between components that are recognizable and generally identifiable as individual grains with a normal optical microscope and those that are below the resolution of the microscope. There is also a distinction between mineral and organic material. The basic components are therefore discussed under three headings: (i) coarse mineral components, (ii) fine components, and (iii) organic components. Taking into account the thickness of the section and the normal resolution of a polarizing microscope, the limit between coarse- and fine-sized basic components will be put at 10 μm for this discussion; however, this does not mean that 10 μm should be taken always as the limit between coarse and fine in soil descriptions (e.g., in c/f-related distribution), as explained in Section 7.2.1. For more information on basic constituents see also Stoops (2015 b) and Stoops and Mees (2018).

6.2. COARSE MINERAL COMPONENTS

6.2.1. Introduction

6.2.1.1. Importance of its Description

The study of the coarse mineral material of a soil can yield much valuable information, including (i) the origin and nature of the parent material, (ii) the profile homogeneity, (iii) the degree of weathering, (iv) the reserve of nutritive elements (presence of large amounts of weatherable components), (v) the natural environment (e.g., phytoliths, biocalcite), and (vi) human influences (e.g., brick fragments and shards). The coarse components of pedogenic origin (e.g., weathering products, pedogenic

Guidelines for Analysis and Description of Soil and Regolith Thin Sections, Second Edition. Georges Stoops.
© 2021 Soil Science Society of America, Inc. Published 2021 by John Wiley & Sons, Inc.
doi:10.2136/guidelinesforanalysis2

minerals) yield information on present and past processes. The main criteria used for the description of the coarse components are nature (composition), size, shape, and internal characteristics. The importance of the study of the coarse fraction of the soil for fertility has been emphasized recently by Small et al. (2014).

6.2.1.2. Subdivision

Four main groups of coarse particles can be distinguished according to their composition and origin: (i) *single mineral grains*, (ii) *compound mineral grains* (e.g., rock fragments, coarse crystalline materials in pedofeatures), (iii) *inorganic residues of biological origin* (e.g., phytoliths, shell fragments, bone fragments), and (iv) *anthropogenic elements* (e.g., bricks, shards, furnace slag, mortars, etc.).

6.2.1.3. Important Properties

6.2.1.3.1. Size

For routine descriptions, the terms silt, sand, (subdivided into very fine, fine, medium, coarse, and very coarse) and gravel are sufficient (see Section 4.3.3 and Table 4.1). It must be remembered that the particle diameters measured in the thin section are not the true diameters of the original particles and the degree of sorting influenced by the random cutting of the grains (see Section 3.1.2).

When the coarse material has an heterogeneous composition, relationships between the nature of the constituents and their size should be investigated. For example, diatoms and phytoliths occur generally in a fine size class, whereas rock fragments can be found mainly in the coarser fractions. In situ weathering may reduce the size of the more weatherable components compared with more stable ones. In alluvial deposits, grains with a higher density are generally smaller than those with a lower density.

6.2.1.3.2. Shape

Shape is a very important characteristic for the genetic interpretation of the coarse basic components of a soil. Euhedral grains are generally of pedogenic origin, formed in situ. Rounded grains of minerals, rock fragments, shards or shells in general indicate transport over some distance, whereas angular fragments are formed by an in situ disaggregation or fracturing, possibly with a transport over short distances (e.g., colluvial transport). Interpretations have to be made carefully, however, as some grains may have been rounded by previous transportation such as grains produced by in situ disaggregation of sandstone. The presence of grains with corroded surfaces, sometimes partly surrounded by dissolution pores, suggests an in situ chemical weathering (Plate 6.1e and f). For the determination of most inorganic residues of biological origin, shape is a characteristic of utmost importance.

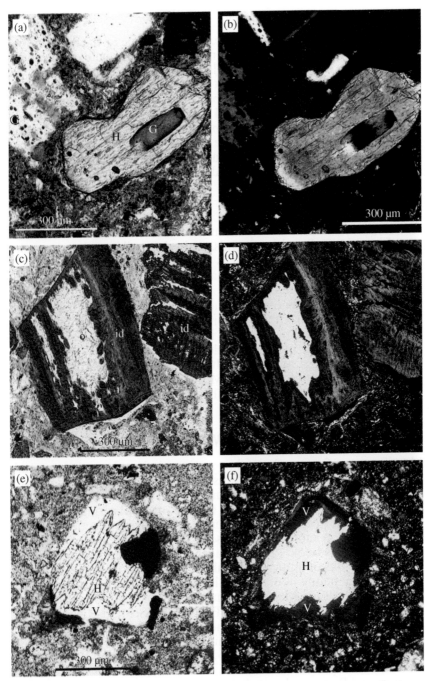

Plate 6.1. *Inclusions and pellicular alteration.* (**a**) hornblende (H) and feldspar (F) grains with inclusions of brownish basaltic glass (G) pointing to a volcanic origin (PPL); (**b**) same, notice the optical isotropic glass inclusions and the undifferentiated b-fabric of the micromass (XPL); (**c**) alteration of olivine (o) to iddingsite (id): Linear and mainly pellicular pattern (left), organized residues (right) (PPL); (**d**) same (XPL); (**e**) pellicular alteration of hypersthene (H) by congruent dissolution; notice serrate basal faces of the crystal and contact voids (v) (PPL); (**f**) same (XPL).

6.2.1.3.3. Abundance and Frequency

Although description of the nature, size, and shape of the different basic constituents of the soil is very important, an estimate of the abundance of these constituents can yield useful information in some cases. For instance, frequency determinations of the amount of weatherable minerals in the coarse fraction are particularly useful for estimating the degree of evolution of the material and its potential fertility. It should be clearly indicated whether the abundance or frequency is expressed on the total area of the slide, on the solid phase, or on the total number of coarse constituents. Rare minerals can easily be overestimated if no absolute count is made. The abundance classes given in Section 4.3.5 should be used.

6.2.1.3.4. Alteration

The term alteration is used here in a very broad sense to include all transformations of rocks, minerals, and other soil constituents under climatic influences (weathering) as well as those caused by geological factors such as injection of magmatic fluids (alteration sensu stricto) and human inpact (e.g., fire).

Although soil scientists are mainly interested in weathering processes and derived products, the products of deuteric and hydrothermal alteration frequently observed in soil thin sections (e.g., iddingsite pseudomorphs after olivine) should be recognized and described by micromorphologists, since they influence the further weathering of the mineral and release of nutrients. Moreover, they can help deduce the nature of the parent material. In this book only a few general guidelines are given.

6.2.2. Single Mineral Grains

Identification of minerals in thin sections is based on their crystallographic characteristics (e.g., crystal faces, cleavage, twinning) and general optical properties. To determine these characteristics, the investigator should be acquainted with the use of a polarizing microscope and the optical properties of minerals. A short introduction in given in Appendix 1. A more detailed discussion of this technique is beyond the scope of this book, and the reader is referred to the numerous excellent textbooks on optical mineralogy mentioned in Appendix 1. For students, very handy books and atlases for the determination of rock forming minerals are MacKenzie and Guilford (1980), MacKenzie and Adams (1993), Melgarejo (1997), Pichler and Schmitt-Riegraf (1997) and Perkins and Henke (2000). Some of the most detailed and complete handbooks for the optical determinations of minerals are Winchell, (1962), Winchell and Winchell (1951), and Tröger (1969, 1971). For information on special methods the reader is referred to Section 3.2 and to Drees and Ransom (1994). A discussion of the interpretation of most common coarse components in soil thin sections, their characteristics, occurrence and alteration, is given by Stoops (2015b) and Stoops and Mees (2018), including many micrographs. Coarse mineral components of pedogenic origin (mainly in pedofeatures) are discussed in several chapters of Stoops et al. (2010, 2018).

The optical properties of mineral grains are usually not mentioned in micromorphological descriptions, since they only serve to identify the mineral. However, some characteristics, such as color, twinning, inclusions, and fluorescence may be recorded if they are unusual or if they concern pedogenic minerals described for the first time in soils. Unambiguous determinations are often difficult in thin sections, such as in the case of carbonate minerals. Several special optical techniques are therefore described in Section 3.2. In addition, several microphysical analyses such as WDXRA, EDEXA, micro-XRD, and micro-FTIRA, have proved to be very useful.

Minerals are often described as either primary or secondary. To geologists, the primary minerals are those developed during the formation of igneous rocks and secondary minerals are those developed after rock formation or belonging to a second generation, while soil scientists consider primary minerals as those inherited from the parent material, with secondary minerals being the pedogenic ones. This may give rise to misunderstanding. For example, epidote formed as a deuteric product is considered as a secondary mineral by a geologist, whereas it will surely be a primary mineral for the soil scientist. Therefore use of the terms primary and secondary is discouraged, and terms such as inherited, detrital, or *lithogenic* vs. *pedogenic* are recommended. A sand composed of different minerals (as in mostly sediments) is often said to be *polymyctic*.

Minerals completely bounded by their own crystal faces are called *euhedral* (or idiomorphic); if they are bounded only partly by their own faces, they are called *subhedral* (or hypidiomorphic); if no crystal faces are present they are said to be *anhedral* (or xenomorphic). As these terms as such add very little information, it is necessary to mention either the actual shape, which is best described in terms of habits or crystallographic forms (e.g., octahedral, bipyramidal or the Miller indices) or the degree of roundness. Because only two of the three dimensions are visible in thin section, a large number of grains with different orientations are needed to determine the habit correctly (see also Section 3.1.2).

Euhedral mineral grains in the groundmass or in voids are generally neoformed (pedogenic). Their crystal habit is commonly an expression of the physicochemical conditions in the soil. For example, gypsum occurs as lenticular crystals in arid soils but has a prismatic habit in most acid sulfate soils.

A description of the internal fabric of detrital mineral grains is important for relating soil parent material to a given rock type. For example, the presence of inclusions of a given mineral (e.g., rutile, sericite) in quartz grains can be useful for identifying the rock from which the grains were derived (Plate 6.1a and b). Other characteristics that should be described include twinning (e.g., in feldspars), zonation (e.g., in feldspars and pyroxenes, pointing to igneous rocks), color, and undulose or wavy extinction (e.g., commonly observed in quartz grains of metamorphic or deformed plutonic rocks). This applies to a primary rock source, but is also valid in the case of polycyclic sediments.

Internal characteristics are especially important in the case of neoformed (pedogenic) mineral grains. Inclusions of other minerals can provide information about the soil material at the time of crystallization. For example, the inclusion of celestite needles in gypsum crystals indicates that the gypsum was formed after the celestite. The presence of dust lines, that are distributions of very fine inclusions in zones parallel to the crystal faces, point to different periods of growth or variations in growth rate of the crystal. Sometimes these features are more distinct using special techniques (e.g., UVF, CL) (Plate 3.4e and f).

Data on mineral alteration provide information both on the environment (type of minerals formed) and on the degree of weathering (amount of alteration products). Interpretation should be made with care, particularly because alteration products may be absent (congruent dissolution) or morphologically unrecognizable.

Mineral grains that have been fractured by natural process may be difficult to identify in thin sections, as grinding during preparation may import artificial fractures. Hiemstra and van der Meer (1997) therefore propose the following criteria as evidence of naturally fractured grains:

1. The grain fragments are juxtaposed and edges can be matched.

2. Some degree of displacement (slip) or rotation is seen.

3. In matrix-supported materials the presence of fine material in the fracture is sufficient evidence.

Schnütgen and Späth (1983) distinguish six principal types of splitting of quartz grains through precipitation of Fe components within the cracks in tropical soils: (i) smooth fracture, (ii) edge fracture, (iii) spheroidal fracture, (iv) triangular fracture, (v) fissure, and (vi) triangular fracture with fissure (Fig. 6.1). They are at the origin of so called

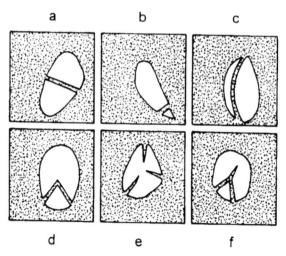

Fig. 6.1. Types of splitting of quartz grains (after Schnütgen & Späth, 1983), (**a**) smooth fracture, (**b**) edge fracture, (**c**) spheroidal fracture, (**d**) triangular fracture, (**e**) fissure, and (**f**) triangular fracture with fissure.

runiquartz grains, typical for lateritic environments (Eswaran et al., 1975).

According to Delvigne (1998) the following morphological and mineralogical features have to be considered when describing the alteration of an individual mineral, both in rocks and in the groundmass:

1. Extent of alteration, expressed either as the percentage of original material that disappeared, or as percentage of newly formed mineral including the voids within that material.

2. Pattern of alteration, that is, the pathway along which alteration progresses through the mineral.

3. Shape, size, and pattern of the residual fragments.

4. Mineralogical composition of the original grain and that of the neoformed material, including changes in color, pleochroism, interference colors, optical orientation.

5. Possible presence of several generations of secondary products, pointing to more complex (polygenetic) evolutions and changing environments.

The first three items are considered in a scheme for the description of the alteration of mineral grains proposed by Stoops et al. (1979). The scheme is valid as long as the original outline of the mineral remains recognizable and the created pore space is either preserved as such or filled with neoformed material. In geology, such features are called pseudomorphs when euhedral grains with clear crystal shapes are concerned. As most grains in rocks are anhedral, lacking clear crystallographic outlines, the term pseudomorph is not applicable in many rock weathering studies. Delvigne (1994) proposed the term *alteromorph* for all cases of transformations of primary minerals (including congruent dissolution) with preservation of the original outlines. When the alteromorph has the same shape and size as the original mineral, alteration is said to be *isomorphous*. When the shape is preserved, but sizes have changed in one, two or three dimensions (e.g., expansion of biotite along the c-axis), alteration is said to be *mesomorphous*. When the shape is completely lost, the alteration is *katamorphous*. In the case of isomorphous or mesomorphous alteration five different *patterns* and five *degrees* of alteration are distinguished (see Fig. 6.2 and Table 6.1), based on an estimation of the volume percentage of the original mineral that has been transformed or dissolved: Class 0 (0–2.5% (i.e., original, almost unaltered material), Class 1 (2.5–25%), Class 2 (25–75%); Class 3 (75–97.5%) and Class 4 (97.5–100%, i.e., completely altered material).

The following patterns are distinguished (Fig. 6.2):

Pellicular: alteration starting at the surface of the mineral and proceeding toward its center (Plate 6.1c through f, Plate 6.2a and b). A core of original material is left but tends to disappear as alteration continues. This type of weathering is commonly found in minerals without well-expressed cleavage and with little or no fracturing (e.g., olivine, Plate 6.1.c and d), but it can also occur as the first alteration stage of minerals with cleavage (e.g., augite, gypsum) or inside a boxwork (see later) (Plate 6.3c and d). Wedging effects on the border of the grains may be mistaken for

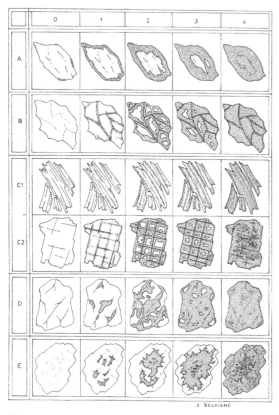

J. DELVIGNE

Fig. 6.2. Patterns of mineral weathering and related terminology (from Stoops et al., 1979). For related terminology see Table 6.1 next page

Table 6.1. Terminology for different patterns and degrees of alteration shown in Fig. 6.1.

Alteration pattern	Degree of alteration (class and volume percentage)				
	0	1	2	3	4
	0–2.5%	2.5–25%	25–75%	75–97.5%	97.5–100%
A		Pellicular	Thick pellicular Large core	Core	
B		Irregular linear	Irregular banded Random residues	Random minute residues	
C.1	Original material	Parallel linear	Parallel banded Organized residues	Organized minute residues	Completely altered material
C.2		Cross-linear	Cross-banded Organized residues	Organized minute residues	
D		Dotted	Patchy Cavernous residue	Dispersed minute residues	
E		Complex			

pellicular weathering (see Section 3.1.3). Grains with pellicular weathering will show a higher degree of alteration the more the section through the grain is removed from the center of the grain, leading to an overall overestimation of the alteration.

Parallel linear: alteration starting along parallel lines inside the mineral, which broaden toward the exterior of the grain. When alteration has progressed (Class 3), only small residues of the original mineral are left, arranged more or less according to a regular pattern (Plate 6.2c through f).

Cross-linear: alteration related to a pattern of intersecting sets of parallel lines. With increasing alteration a **cross-banded** pattern develops.

Parallel and cross-linear patterns are commonly observed in minerals with well-expressed cleavages (e.g., amphiboles, pyroxenes, and micas). The orientation of the mineral in thin section may influence the pattern observed. For example, inosilicates that are cut perpendicular to their c-axis exhibit a cross linear pattern, whereas a section parallel to the c-axis shows a parallel linear pattern. Linear patterns may also be caused by twinning (plagioclase, microcline) or exsolution (e.g., perthites). (Plate 6.2c and d).

Irregular linear: alteration related to an irregular pattern of lines. In the last stages of alteration, small isolated residues are dispersed at random in the alteration products. This pattern is commonly found in minerals without pronounced cleavages but several fractures. Examples are olivine, garnet, or quartz (Plate 6.3a through d).

Dotted: alteration starting within isolated points (e.g., inclusions) and progressively hollowing the grain. This pattern is less common. One should realize that the dotted pattern in two-dimensions might actually result from fingerlike embayements in three-dimensions (Plate 6.3e and f).

For each different degree and pattern of alteration a specific name is proposed: for Class 1, the terms refer to the morphology of the secondary products and/or created pore spaces, and for Class 3, to the morphology of the residual part of the mineral (Table 6.1). For Class 2, both characteristics can be used.

Combinations of two or more of these patterns are frequent. When alteration reaches Class 3, the different alteration patterns are not always distinguishable.

In addition to the alteration patterns mentioned above, which result mainly from chemical processes, tubular alteration due to biological activity (Plate 6.4a and b) also has to be considered in soils. Jongmans et al. (1997) described the presence of small (20- 30 μm) channels caused by fungal activity in feldspar grains from albic horizons in Spodosols. The channels are characterized by a constant diameter. Tubular alteration of volcanic glass by bacteria was described by McLoughlin et al. (2008 and 2010) in marine environment, and recently by Burns et al. (2017) in volcanic ash soils. It are tubular pores of a few micrometers in diameter and tens to hundreds

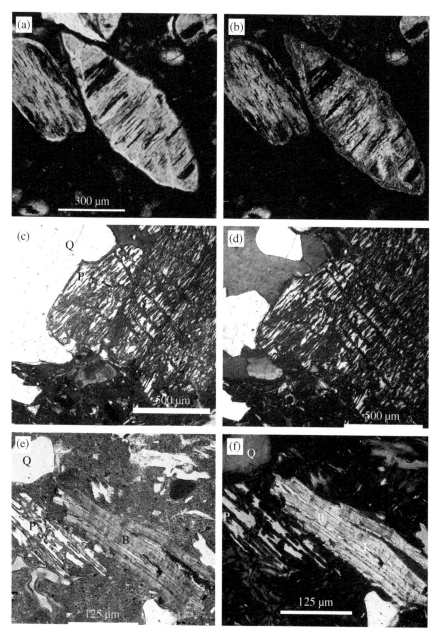

Plate 6.2. Alteration patterns of minerals (**a**) pellicular alteration of lenticular gypsum to bassanite forming pseudomorphs; notice persistence of original cleavage (NE-SW in the central crystal) (PPL); (**b**) same, dehydration is clearly stronger at the border (b) of the crystals (higher interference colors) than in their center (XPL); (**c**) parallel linear alteration of perthite (P); the voids created by the dissolution of the albite lamellae are filled with yellowish clay, at right a large quartz grain (Q) (PPL); (**d**) same, notice that the quartz grain is in fact a quartz aggregate, and that the yellowish clay is present as clay coatings (XPL); (**e**) parallel linear alteration of biotite (B) in yellowish homogeneous micromass; oriented fragments of altered perthite (P) and infillings of strongly oriented fine yellowish clay (ci), quartz grains (Q) (PPL); (**f**) same, notice interlamellar deposits of parallel oriented kaolinite (K) in the biotite, and presence of mica flakes (M) in the groundmass (XPL).

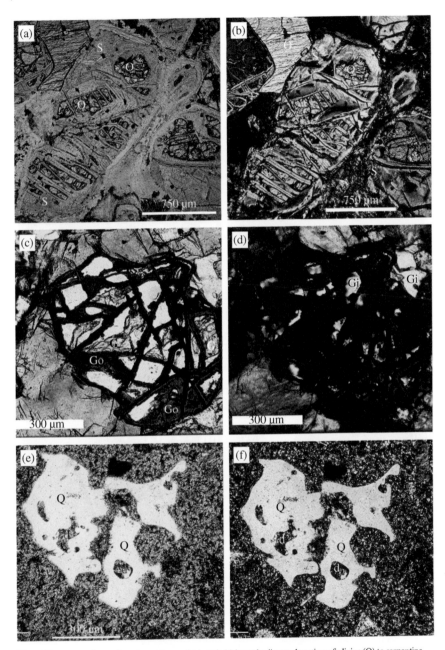

Plate 6.3. Irregular linear alteration patterns of minerals (**a**) irregular linear alteration of olivine (O) to serpentine (S); several steps can be recognized (PPL); (**b**) same, notice that fragments belonging to the same original olivine crystal have identical interference colors, showing their "optical continuity" (XPL); (**c**) irregular linear alteration of garnet (G) with formation of a goethite (Go) boxwork (dark brown), by congruent dissolution of the garnet fragments, contact voids are formed (v), notice high relief of garnet (PPL); (**d**) same, some of the contact voids are filled with coarse gibbsite (Gi) (white) (XPL); (**e**) dotted alteration of quartz (Q) in rhyolite; notice that some of the dots (d) may me the result of deep embayements in the third dimension (PPL); (**f**) same (XPL).

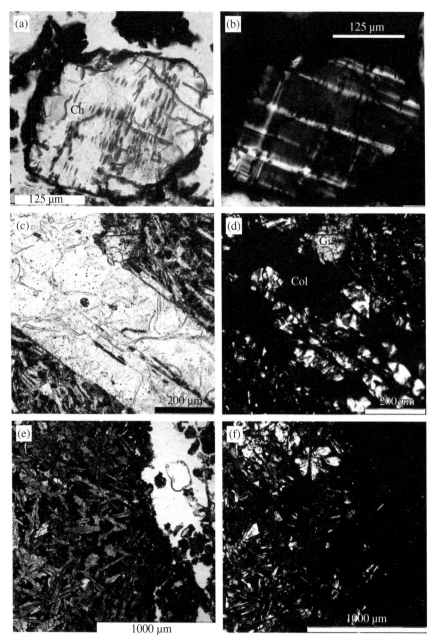

Plate 6.4. *Alteration of minerals and rocks* (**a**) tubular alteration of a microcline grain; notice channel shaped voids (ch) in grain, recognizable by their rounded endings (PPL); (**b**) same, notice the cross-hatched or "tartan" twining of microcline (XPL); (**c**) polygenetic alteration of feldspar phenocrystal in basalt fragment (PPL); (**d**) same, notice the presence of isotropic colloids (Col), which are partly transformed to gibbsite (Gi) (XPL); (**e**) pellicular alteration of basalt gravel in soil; notice brown impregnation on the right side of the fragment (PPL); (**f**) same, notice presence of small fresh feldspar crystals in the unaltered lower left corner, the isotropic aspect of most of the altered part, and the gibbsite (Gi) in the upper altered part (XPL).

micrometers long that can be distinguished from abiotic a ray recoil trails (so called fission tracks) that are a magnitude smaller than the biotic tunnels. In sedimentology, the process was described on a larger scale (e.g., Tudhope and Risk, 1985) (see also Plate 6.9e).

Minerals sometimes undergo a generalized or homogeneous alteration, which is the transformation is not localized in preferential zones of the mineral but starts and proceeds quite homogeneously throughout the whole volume of the grain (e.g., sericitisation of feldspars, bleaching of biotite).

In several cases of alteration the mineral is not replaced by pedogenic products but rather undergoes congruent dissolution and an empty space is left. This has been observed in weathering of calcite, quartz, and sometimes pyroxenes, amphiboles, and garnet. This form of alteration (mostly of the pellicular type) can be observed only when the resultant moldic void is preserved (Plate 6.1e and f).

As mentioned above, the most important characteristics to be described are the pattern and degree of alteration and the nature (composition) of the pedogenic products. The fabric of the pedogenic products should also be described, especially in the case of medium- and coarse-grained products (e.g., grain size, shape, and referred orientation pattern). A few examples of description of altered mineral grains are given in Delvigne (1998), Stoops (2015b) and Stoops and Mees (2018). As a result of pedoplasmation, the fabric of the alteration product can be transformed (Stoops and Schaefer, 2018)

Alteration is often a polygenetic or polyphased process, whereby subsequently other products are formed or the initially formed alteration products are destabilized and transformed to new phases. This is for instance the case for biotite weathering to vermiculite and subsequently to kaolinite, or for plagioclase weathering to a Si-Al-colloid, that later transforms to gibbsite by further leaching (Plate 6.4c and d). In such a case, the end result of the first alteration step will be considered as Class 0 for the second alteration sequence. For more details and ample examples the reader is referred to Delvigne (1998). Good examples for the alteration of olivine are given by Delvigne et al. (1979) and for the alteration of biotite by Bisdom et al. (1982).

A common case of complex alteration sequence patterns is that of the formation of *boxworks* (Plate 6.3c). Alteration proceeding according to a linear pattern forms *septa* of relative stable minerals (e.g., coarse goethite). Changes in the microenvironment around the mineral leads to a gradual, mostly pellicular congruent dissolution of the remaining primary mineral fragments. The space created between the septa and the fragments is called a *contact void* (Nahon, 1991). The resulting feature is called a boxwork. The voids may remain empty or become filled with neoformed minerals. They may be of the same nature as the septa (e.g., goethite), but they are then generally cryptocrystalline, or of different nature (e.g., gibbsite), either formed in situ or translocated in suspension (Plate 6.3c and d).

Delvigne (1994, 1998) elaborated a comprehensive terminology to name and classify the different types of alteromorphs, based on their morphology and porosity. For more details the reader is referred to the original paper and atlas.

6.2.3. Compound Mineral Grains and Rock Fragments

There is no simple key for the determination and identification of rock fragments. A basic knowledge of petrography and experience are required for accurate identifications. The reader is referred to manuals on petrography for precise determination of rock fragments. For igneous and metamorphic rock fragments, the nomenclature of the subcommission of the International Union of Geologists should be used by preference (Le Maitre, 2002; Fettes and Desmons, 2007); for sedimentary rocks widely used systems are proposed by Dunham (1962), Folk (1962), Tucker, (2001), Boggs, (2009), Flügel, 2010 and Pettijohn et al. (2012). Easy to use atlases are also available, including MacKenzie et al. (1982), Adams et al. (1984), Yardley et al. (1990), MacKenzie and Adams (1993), Adams and MacKenzie (1998), and Scholle and Ulmer-Scholle, 2003.

If no precise petrographic determination of rock fragments can be given, a description, as precisely as possible, should be presented. This should at least include the nature of the main components, size of components (e.g., micro- or macrocrystalline (see Section 4.3.3), their fabric, and whether the rocks are plutonic, volcanic, metamorphic, or sedimentary.

Small rock fragments of relatively coarse-grained material are difficult to identify in soil thin sections, as only a limited, non-representative number of constituents are visible, and some of them might be selectively weathered. For example, granite, gneiss, micaschist, or quartzite may yield similar aggregates consisting only of quartz (e.g., Plate 6.2c and d). The same is true for coarse-grained limestone, marble, carbonatites, or calcite infillings of igneous rocks. Therefore fragments consisting of a single mineral species (monomineralic) should be named according to their composition: for example "quartz aggregate" instead of "quartzite fragment". For polymineralic fragments of igneous rocks, the petrographic term with the suffix '-oid' is proposed to indicate specific combinations of minerals, irrespective of their actual genesis. Due to selective alteration, the composition of rock fragments can be different from that of the rock from which they derived. The key in Table 6.2 is therefore not meant as an alternative for petrographic determinations, but only as a first help for students with insufficient background in geology. Appropriate terms to describe the internal fabric of rock fragments can be found in the manuals on petrography mentioned before.

The color and relief of volcanic glass should be specified. A light color and negative relief points to a Si-oversaturated (acid) glass; a brownish color and positive relief to a neutral or undersaturated (basic) glass. Guidelines for the description of volcanic material in soil thin sections are given by De Paepe and Stoops (2007) and Stoops et al. (2018c).

Table 6.2 Key to the determination of some of the most common rock fragments.

1. The dominant constituent of the fragment is volcanic glass:
  1.1. The glass has a brownish color, positive relief and enclosed minerals are mainly one or more of the following: Ca-rich feldspars, pyroxenes, olivines and opaques. Vacuoles may be common:
  basaltic glass
  1.2. The glass has an egg-yellow to orange color, a negative relief, and enclosed minerals and vacuoles as above:
  palagonite
  1.3. The glass is practically colorless, with negative relief. Enclosed minerals are alkali-rich feldspars, corroded quartz, ores; vacuoles may be common:
  rhyolitic glass
  1.4. As above, but with numerous vacuoles, resulting in a foam like appearance:
  pumice
2. Euhedral or subhedral silicate crystals dominate in a microcrystalline and/or glassy matrix:
  2.1. Mineralogical composition as 1.1:
  basaltoid fragment
  2.2. Mineralogical composition as 1.3:
  rhyolitoid fragment
3. The fragment consists of medium to coarse, interlocking minerals without cement or matrix:
  3.1. Only one mineral type is present: **monomineralic aggregates**, named according to composition, e.g. aggregates of quartz, pyroxenes, amphiboles, olivine, serpentine, calcite, gypsum;
  3.2. Different silicate minerals without pronounced parallel orientation are present:
  3.2.1. Quartz, alkali feldspars, micas (muscovite and/or biotite) and/or amphiboles:
  granitoid
  3.2.2. Alkali-feldspars and alkali-amphiboles or-pyroxenes, (+ quartz):
  syenitoid
  3.2.3. Alkali-plagioclases, amphiboles, (+ quartz):
  dioritoid
  3.2.4. Ca-plagioclases, pyroxenes, ores, (+ olivine):
  gabbroid
  3.2.5. feldspars and feldspathoids (+ pyroxenes):
  foidoid
  3.3. Different silicate minerals with a pronounced parallel orientation are present:
  3.3.1. Alternating layers of feldspar (+ quartz) and mica (+ hornblende):
  gneiss
  3.3.2. Relative large mica plates, some small feldspar and/or quartz grains:
  micaschist
  3.3.3. Amphiboles, plagioclases, epidotes, zoisites:
  greenstone
4. The rock fragment consist of small sericite and/or chlorite flakes with very fine grains of quartz:
  sericite or chlorite schist
5. The rock fragment consists of a more or less dense packing of clastic angular or rounded, dominantly silicate grains with a cement or a matrix:
  5.1. Clastic grains > 2000 μm:
  5.1.1. The grains are angular:
  breccia
  5.1.2. The grains are rounded:
  conglomerate
  5.2. Clastic grains 63- 2000 μm:
  sandstone
  5.3. Grains < 63 μm:
  siltstone or mudstone
  (Classify further according to composition of grains (e.g. quartz, feldspar, glauconite) and type of cement, e.g,. calcareous, siliceous, ferruginous).
6. The rock fragment consists dominantly of calcitic elements, including fossils:
  limestone

The description of altered rock fragments is best approached through an analysis of the pores that have been created, and whether the voids are empty or filled with neoformed products. The related distribution of the voids with respect to the individual grains is particularly important. Three main types are distinguished (Bisdom, 1967; Stoops et al., 1979): *inter-, intra-* and *trans*mineral voids. Transmineral voids traverse the rock without following the grain boundaries; intermineral voids follow the grain boundaries, and intramineral voids occur within a single mineral grain and their pattern is often related to specific crystallographic directions. For the other characteristics of the voids and void patterns, the relevant criteria listed in Section 5.2.2. should be followed.

No general system for the description of altered rocks fragments has yet been established, but the following guidelines may be helpful. With respect to the pattern of alteration, the same terms as for mineral alteration are applicable in many cases (e.g., pellicular alteration of basalt fragments (Plate 6.4e and f), linear alteration of schist fragments). The degree of alteration of a rock is difficult to define for the following reasons:

1. Alteration does not proceed homogeneously in a rock.

2. The position of the section will influence the visible degree of alteration, especially in the case of pellicular alteration; e.g., a section through the center will probably show a thin alteration rim, whereas a tangential section near the border will give the impression of a totally altered material.

3. Easily weatherable minerals will be weathered, whereas very stable minerals remain and accumulate.

4. Cements can have a stability that is quite different from that of the detrital components.

5. Both physical and chemical alteration can occur.

For example, a granite fragment may be completely chemically altered, with the rock fabric still clearly recognizable, or the rock may have mechanically disintegrated completely, although all minerals seem fresh. Generally, some mineral species will be altered and others will show little or no alteration. In addition to void pattern, alteration of the mineral grains and the nature of the neoformed products present in the voids should be described.

The following scheme is proposed for the description of altered rock fragments:

General pattern of alteration: terms proposed for the alteration patterns of mineral grains (e.g., pellicular, parallel linear) in addition to tubular and homogeneous alteration patterns.

Void pattern in the altered zone and the nature of void infillings, if any.

Weathered minerals and the neoformed products; attention should also be given to the alteration of cements in the case of clastic rocks.

For more detailed studies of altering rock fragments, differences between different altered zones (e.g., concentric bands in the case of pellicular alteration) and their boundaries (regularity and sharpness) should be considered.

A detailed description of the weathering morphology of each mineral species and their correlation in each alteration zone is necessary. Examples of rock alteration are discussed for instance in Delvigne (1998), Zauyah et al. (2018) and its degradation in Stoops and Schaefer (2018).

6.2.4. Inorganic Residues of Biological Origin

6.2.4.1 Introduction

Inorganic residues of organic origin (composed mainly of opal, calcite, or phosphate) comprise phytoliths, diatoms, fragments of the internal or external skeletons of animals, or mineral products of animal metabolism. Some are discussed briefly below. For more detailed information, the reader is referred to Simkiss and Wilbur (1989) and chapters in Nicosia and Stoops (2017) and Stoops et al. (2018c). Fresh or decomposing plant or animal tissues are treated in Section 6.4.

6.2.4.2 Opaline Components

Opaline particles can easily be recognized in thin section by their moderate negative relief and optical isotropism; they are mostly colorless and hyaline. Shapeless opaline particles can easily be confused with acid volcanic glass. Occurrence and interpretation of biogenic opaline constituents are discussed by Kaczorek et al. (2018).

Opal phytoliths: small (commonly 5–250 µm) opal bodies, formed in the lumen of plant cells, in intercellular spaces, or in cell walls of Pteridophytes, Gymnosperms, Monocotyledons and Dicotyledons. Grasses, evergreens, and equisetaceous plants have a particularly high Si-content. They are usually transparent and contain fine inclusions of carbon that produce a brown or purple color, especially after burning. In PPL they have a distinct negative relief, in OIL or TDFI they generally show a clear whitish color (Plate 3.3c). Some phytoliths show in UVF or BLF autofluorescence, e.g., in Spodosols and Andisols (Van Vliet-Lanoë, 1980). This is generally not the case for recently formed individuals (Altmüller and Van Vliet-Lanoë, 1990). The shape and size of phytoliths are function of vegetation type, and in particular of the plant part from which they are derived. Phytolith deposits were described earlier in 1936 by Lacroix (Deflandre, 1963), who called them *mascareignite*. The first microscopic descriptions were recorded by Ehrenberg in 1846 (Deflandre, 1963) naming them Phytolitharien. For further detailed information the reader is referred to Piperno (1988) and Madella and Lancelotti (2012). Specialists distinguish specific morphologies, typical for specific plant tissues and families, and less specific or redundant morphologies (Rovner, 1971). For the occurrence and the role of phytoliths as soil constituents see Drees et al. (1989), Golyeva (2008), Vrydaghs et al. (2017) and Kaczorek et al. (2018) and the references therein.

Several classification schemes for phytoliths have been proposed (e.g., Twiss et al., 1969; Piperno, 1988; Runge, 1999), but they are all based on three-dimensional observations, and therefore not directly applicable to specimens observed in thin sections. The aim of the description criteria proposed below is to make comparison between materials easier and not to identify the phytoliths. The problems of observation of phytoliths in soil thin sections are discussed by Vrydaghs and Devos (2018),

Table 6.3 and Fig. 6.3 present a tentative list of two-dimensional shapes of opaline phytoliths.

The elongated shapes can be moreover bilobate or polylobate. A more detailed two-dimensional scheme is given in ICPN 0.2 (2019). In addition to shape, the surface roughness should be described by adding the terms *serrated* or *spiny* when required (Plate 6.5c). Apart from shape, the presence and distribution (dispersed or central) of inclusions should be noted.

Little information is available on the alteration of phytoliths. Apart from mechanical fractures, chemical corrosion has been mentioned by Benayas Casares (1963) and Jenkins (2009). Heating above 800 °C, for instance in wood fires, results in melting, forming lumps of vesicular glass (Runge, 1998; Brochier, 2002; Canti, 2003) generally showing UVF or BLF (Gebhardt and Langhor, 1993; Devos et al., 2009).

Table 6.3. Tentative subdivision of two-dimensional shapes of opaline phytoliths.

Angular	Square
	Rectangular
	Elongate
	Triangular or fan-shaped
	Polyhedral
	Other
Rounded	Circular
	Oval
	Oblong

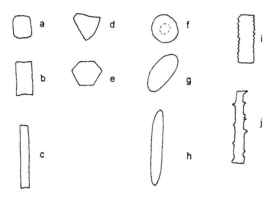

Fig. 6.3. Some shapes of phytoliths: (**a**) square, (**b**) rectangular, (**c**) elongate, (**d**) triangular or fan-shaped, (**e**) polyhedral, (**f**) circular with central depression,(**g**) oval, (**h**) oblong, (**I**) serrated, and (**j**) spiny.

Phytoliths generally occur in the groundmass (Plate 6.5a and b), but they may also occur as infillings (Plate 6.5e and f). In the groundmass, four types of distribution patterns can be distinguished: articulated (preserving the original pattern they had in the plant remain, pointing to an in situ decomposition, not followed by pedoturbation) (Plate 6.5b and d), and three types of disarticulated and well separated distributions: isolated, clustered, and at random. Whereas the three first patterns are dominant in archaeological materials (Vrydaghs et al., 2017; Kackzorek et al., 2018) the last one is dominant in soils.,

Diatoms: unicellular algae with an external skeleton of opal (frustules); they are colorless and have a distinct negative relief. They can have a radial or bilateral symmetry and assume many different shapes. Diatoms can be inherited from the parent material (e.g,. marine) or be remnants of soil microflora especially in very wet soils, such as temporary flooded alluvial soils and paddy fields (Plate 6.6 a through d) or can be incorporated in the topsoil as part of herbivore excrements (Brönnimann et al., 2017a). For more details see also Verleyen et al. (2017) and Kackzorek et al. (2018).

Radiolaria: the opaline outer skeleton of these unicellular organisms is frequently found in marine sediments. They have a more or less spherical shape composed of a number of architectural elements, yielding a sieve-like porous morphology and are always inherited. For more details see Verleyen et al. (2017) and Kackzorek et al. (2018).

Sponge spicules: colorless cylindrical opaline bodies with a distinct negative relief and a narrow central channel (Plate 6.5b, Plate 6.6e and f). They are frequently observed in soils on marine, lacustrine, or alluvial materials, (inherited) or in situ formed in waterlogged soils and on vegetation in spray zones (Stoops et al., 2001). Sponge spicules have been frequently reported in loess (Wilding and Drees, 1968). For more details see Vrydaghs (2017) and Kackzorek et al. (2018).

6.2.4.3 Calcium Oxalate Crystals

Calcium oxalate phytoliths: common bodies in some plant tissues (Plate 6.7 a through d), they are mostly euhedral or subhedral, and composed of calcium oxalate (whewellite, seldomly weddellite) (Jaillard et al., 1991; Verrecchia, 1990). Generally small oxalate crystals are difficult to distinguish from carbonate crystals by their optical properties alone. Oxalate phytoliths are common in excrements of herbivores and stable deposits (Brönnimann et al., 2017a) and in plant ashes where they are transformed to calcite alteromorphs (Canti and Brochier, 2017).

6.2.4.4. Calcium carbonate constituents of biological origin

Calcite phytoliths: so called cytomorphic calcite grains, optically generally not possible to distinguish from calcium oxalate phytoliths. They are common in roots in semiarid regions (Plate 8.21c through f).

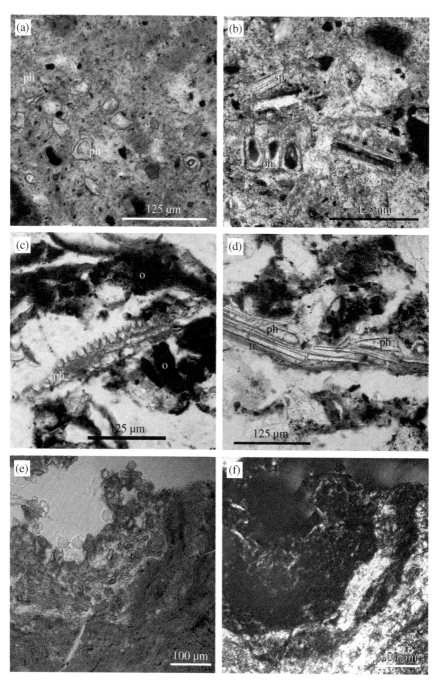

Plate 6.5. *Opaline residues of organic origin.* (**a**) square and circular phytoliths (ph), some with central depression (PPL); (**b**) three interconnected (articulated) rectangular phytoliths (ph) with central diffuse carbon inclusions, above them a sponge spicule (sp) with central canal (PPL); (**c**) spiny, elongate phytoliths (ph) and humified organic remains (o) (PPL); (**d**) interconnected elongate phytoliths (ph) with part of the plant tissue preserved (tissue residue (tr) (PPL) (**e**) loose phytoliths infilling channel with clay coating (PPL); (**f**) same, notice isotropism of phytoliths (XPL).

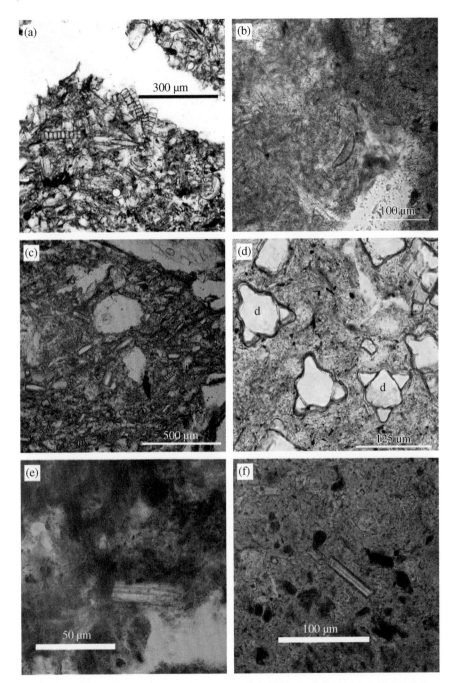

Plate 6.6. *Opaline residues of organic origin.* (a) accumulation of diatoms in paddy soil on lacustrine deposits (PPL); (b) diatoms and phytoliths forming upper horizon of bog-ore soil (PPL) (c) diatoms in bog ore (PPL); (d) diatoms (d) in hydromorphic soil (PPL); (e) sponge spicule; notice central channel (PPL); (f) sponge spicule and phytoliths (PPL).

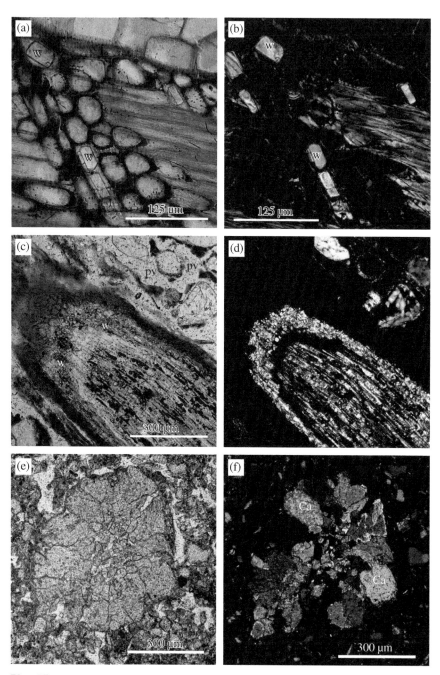

Plate 6.7. *Inorganic residues of organic origin.* (**a**) fresh organ residue (at least three different cell types) with Ca-oxalate (whewellite: W) phytoliths (PPL); (**b**) same, notice strong interference colors of the cell walls, the high interference colors and arrangement of the whewellite (w) phytoliths (XPL); (**c**) oblique section through a root with numerous Ca-oxalate (whewellite: W) crystals; notice the grains of abrasive in the packing voids (pv) (PPL) (**d**) same; the small particles with high interference colors are the whewellite crystals (w); the interference colors of first order gray in the center are due to cell walls (c) (XPL); (**e**) calcite biospheroid excreted by earthworm; notice the smaller grain size in the center (PPL) (**f**) same, notice the higher order interference colors of the large calcite (Ca) and the smaller calcite crystals in the center (XPL).

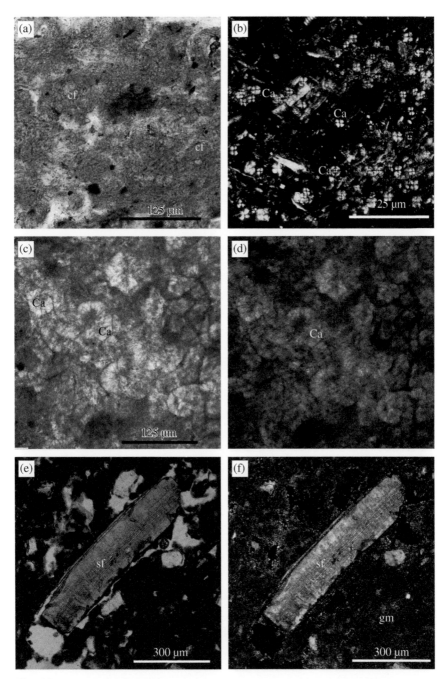

Plate 6.8. *Calcareous residues of organic origin.* (**a**) herbivore excrement composed of cell fragments and amorphous organic matter, with calcareous spherulites (PPL); (**b**) same, notice pseudo uniaxial interference figures in spherulites (Ca) (XPL); (**c**) calcite spherulites (Ca) with fibro-radial fabric in laminar layer of calcrete (PPL); (**d**) same, notice the radial fabric in the calcite spherulites (Ca) (XPL); (**e**) shell fragment (sf) in groundmass with porphyric c/f related distribution pattern and grayish, calcite rich micromass (PPL); (**f**) same, notice high interference colors in shell fragment and calcitic crystallitic b-fabric of groundmass (gm) (XPL).

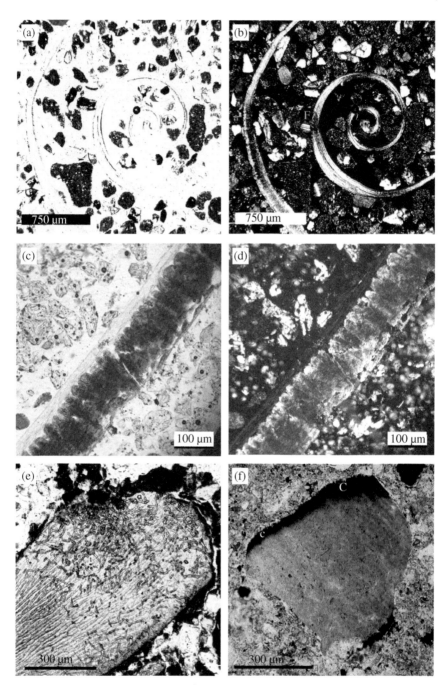

Plate 6.9. *Calcareous residues of organic origin.* (**a**) shell of snail (PPL); (**b**) same (XPL); (**c**) section through an egg shell; notice the transparent layer and the mammillary knobs at left (PPL); (**d**) same, notice the more or less parallel zonation perpendicular to the surface, the high interference colors of the carbonate part and and the isotropic layer on top (XPL); (**e**) fragment of shell showing tubular (t) alteration (PPL); (**f**) bone fragment, partly charred (c) (PPL).

Biospheroids: ovoids of calcite, 0.5 - 2 mm in diameter, composed of an outer, tightly packed layer of coarse (50–250 µm) prismatic calcite crystals with an incomplete radial fabric and finer (20 -60 µm) crystals in the center (Plate 6.7e and f). They are secreted by earthworms (genus Lumbricus) (Bal, 1977, Becze-Deák et al., 1997, Canti, 1998, Canti and Piearce, 2003, Canti and Brochier, 2017a

Calcite spherulites: spherical grains of 5 to 15 µm, usually composed of acicular calcite crystallites arranged in a radial pattern, showing a straight extinction cross between crossed polarizers, called a pseudouniaxial (negative) figure (Plate 6.8a and b). Small spherulites show first order interference colors, whereas the larger ones sometimes show interference colors of second order. They were observed in herbivore excrements in several archaeological deposits (Brochier et al., 1992; Brochier, 1996). Comparative studies (Canti, 1997, 1999) have shown that calcite spherulites are formed in the digestive tracts of herbivores. For more details see Canti and Brochier (2017a).

Spherulites of 0.5 to 100 µm in diameter, with fibroradial fabric, are commonly observed in laminar layers of calcretes (Plate 6.8c and d) and are considered as products of biomineralization by bacteria or cyanobacteria (see review by Verrecchia et al., 1995).

Shells of bivalves (e.g., oysters) and gastropods (e.g., land snails): consist mainly of calcium carbonate (aragonite and/or calcite) and are thus characterized by high interference colors. Nearly parallel elongated crystals form layers with different orientation. In XPL often an iridescent aspect is shown. They can be recognized by their external shape and fibrous internal fabric (Plate 6.8e and f, Plate 6.9a and b). Mollusk shells are inherited from the parent material; shells of snails may be inherited or be remnants of the present or the past fauna. Calcite crystals from slugs may be distributed in the soil directly or via fecal material of their predators (Canti, 1998). Tubular alteration is often observed (Plate 6.9e). A special alteration of mussel-shells to needle shaped calcite was observed by Villagran and Poch (2014). For more details see Canti (2017a and b) and references therein.

Shells of eggs: uncommon in natural soils, but they have been observed in archaeological sites. They are generally elongated features consisting of several layers of calcite crystals. In the outer layer, which is the most important, the columnar crystals are oriented perpendicularly to the surface (the palisade layer) and have a feather-like fabric. A thin (50–75 µm) layer of aragonite spherules (the so called mammillary knobs) occurs at the inner side. As the calcite crystals in the palisade layer are not strictly parallel, but show angles op to 10°, a strippy appearance is observed in XPL (Plate 6.9c and d). For more details see Canti (2017c).

6.2.4.5 Bones and Other Skeletal Tissues

Bones are a combination of fibrils consisting of platy shaped bioapatite crystallites of a few nanometers in size, and collagen fibers. These fibrils can be packed in different ways. Bones have a fibrous internal fabric, a

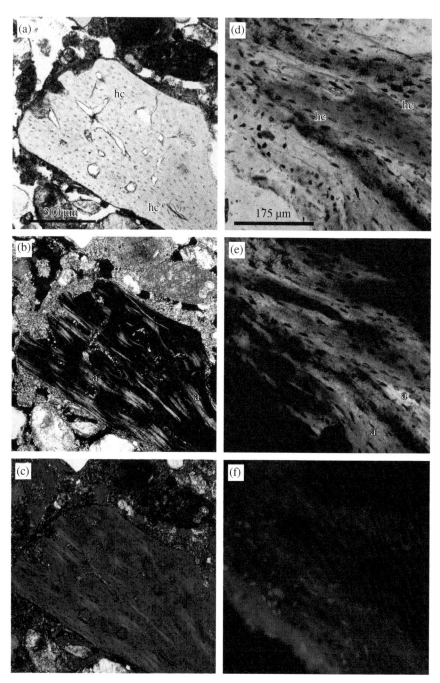

Plate 6.10. *Bone fragments.* (**a**) bone fragment in calcareous soil; notice haversian canals (hc) (PPL); (**b**) same, with clear light gray interference colors of the apatite fibers (a); notice crystallitic b-fabric in groundmass (XPL); (**c**) same, showing the parallel orientation of the apatite fibers, giving a pseudo-length slow aspect to the entire bone (XPLλ) (nγ oriented NW-SE); (**d**) bone fragment; notice haversian canals (black elliptic points) (hc) (PPL); (**e**) same; notice light gray interference colors of the apatite fibers (a) and haversian canals (XPL); (**f**) same, faint green and reddish fluorescence colors (UVF).

light yellow color, first order white to gray interference colors and strong blue UVF and yellowish green BLF (Plate 6.10). Apatite as such is length fast (Plate 6.10c), but in combination with the collagen the fibrils exhibit a length slow orientation due to form birefringence (see Section 4.3.2.2.3). The optical orientation changes depending on the type and degree of alteration, the embedding medium, etc. For instance after heating, the length slow orientation of apatite reappears. A characteristic feature is the presence of abundant small (about 2–5 µm in diameter) channels (haversian canals) (Plate 6.10a, d and e). Bone splinters are commonly observed in excrements of carnivores (Brönnimann et al., 2017b). In archaeological sites bones are quite often partly burned, resulting in brownish rims (Plate 6.9f).

Other skeletal tissues such as teeth, and keratine (horn, hoof) all exhibit white to gray low interference colors and autofluorescence.

For more details on bones and other skeletal tissues, their optical characteristics, and physical and chemical alteration, see Villagran et al. (2017) and Karkanas and Goldberg (2018).

6.2.5. Anthropogenic Elements

The presence of anthropogenic elements in a soil provides proof of human influence. Such elements occur mainly in the upper horizons of cultivated soils and archaeological layers, but they can also occur in the parent material of colluvial and alluvial soils. Most often they have sharp boundaries with the soil matrix, and an internal fabric different from the groundmass. The most common types are briefly described below. For more information, the reader is referred to more specialized papers in archaeological journals and to Nicosia and Stoops (2017) and Macphail and Goldberg (2018).

Ceramics: fragments of pottery, bricks, tiles and drainage pipes, can have different colors in transmitted light, depending on composition and heating: a white or beige color points to an iron-free base clay; an yellow-beige to a calcareous iron bearing clay heated in a oxidizing atmosphere; a red color points to Fe bearing clay baked in oxidizing atmosphere (Plate 6.11a and b); a gray to black color to organic and/or Fe-bearing clay heated in reducing atmosphere.. The characteristics of the fine material in bricks and pottery depend strongly on the firing temperature; if interference colors are still visible in the micromass, the temperature remained below the break-down temperature of the clay; in extreme cases a partial vitrification is observed. The coarse components may yield information about the origin (local vs. imported earthware). For more details see Maritan (2017), Quin (2013), and Betancourt and Peterson (2009).

Furnace slag: a heterogeneous material, partly or totally consisting of a dark isotropic vitreous phase. At a first glance it is easily confused with basic volcanic glass. Crystalline parts show skeletal or dendritic growth of some minerals (e.g., magnetite, ilmenite, fayalite) (Plate 6.11c and d). For more details see Angelini et al. (2017).

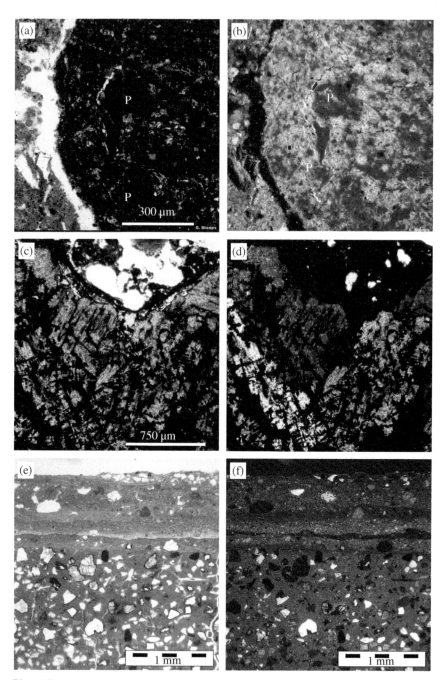

Plate 6.11. *Anthropogenic components.* (**a**) fragment of pottery notice its partially opaque aspect (PPL) (**b**) same, notice typical reddish brown color of the pottery fragment (P), and calcitic crystallitic b-fabric of the groundmass (XPL); (**c**) fragment of furnace slag: Notice inclusions of dendritic opaque mineral (**d**) (PPL) (**d**) same, the interference colors indicate that the slag is composed of irregular, equidimensional coarse mineral grains with inclusions of dendritic opaque minerals (XPL); (**e**) fragment of Roman stucco; notice the different parallel layers of plaster (PPL); (**f**) same; notice the very fine calcitic-crystallitic b-fabric, characteristic for calcareous plasters and mortars (XPL).

Metal fragments: appear as opaque particles in PPL, with a metallic luster in OIL.

Glass fragments: readily recognized by their hyaline aspect and isotropism, but splinters easily confused with acid volcanic glass.

Calcitic mortar and plaster fragments: often observed in archaeological sites. They can be confused with impure limestone fragments. Inclusions of straw, brick or bone fragments and presence of areas of cryptocrystalline calcite may help identifying them. Their porosity is generally higher than that of limestone. In the case of plasters, a banded arrangement of the constituents is generally observed (Plate 6.11e and f). For more details and references see Stoops et al. (2017a).

Gypsitic mortars and plaster fragments: rather rare, because of their solubility. They have a microcrystalline micromass, sometimes transformed to a poikilotopic one in old mortars (Stoops et al., 2017b).

Coal fragments: considered an indication of human influence if no coal seams are present in the parent rock. Four types of coal can be distinguished: lignite, cannel coal, bituminous coal, and anthracite. Whereas lignite has a deep reddish orange to black color in PPL, the other types are opaque in PPL, and black in OIL (Canti, 2017d). For its precise determination a study in reflected light microscopy is necessary (Ligouis, 2017).

Fertilizers: of different compositions, can occur in surface horizons (Stephan, 1972).

For describing the alteration of anthropogenic elements such as shards and brick fragments no commonly used specific terms exist as yet. Alteration of furnace slag, mortars and plaster can be treated the same way as that of rock fragments.

6.3. DESCRIPTION OF FINE MINERAL COMPONENTS

6.3.1. Importance of its Description

The fine soil material, because of its size, has specific physical and chemical characteristics and a particular mechanical behavior. These characteristics are expressed morphologically by features indicating processes such as translocation and reorientation of the fine particles, and absorption of staining substances. In many soils, the fine material is the only pedogenic component. The fabric and composition of the fine material thus reflects quite well the history and evolution of the soil material and its behavior.

The most important characteristics of the fine material are nature, size, shape, frequency, and organization. These can be used to describe both mineral and organomineral fine material.

6.3.2. Nature

All or part of the components of the fine material are generally too small to be recognizable at the scale of observation, and many are beyond the resolving power of the optical microscope at the highest magnification used (see Section 3.1.4). In most cases, the mineralogical composition of the fine material cannot be determined directly and unequivocally by normal optical methods. Moreover, the presence of organic and inorganic amorphous material may complicate such determinations. Therefore, characteristics related to composition and which can be observed by normal microscopic techniques should be described (e.g., color, limpidity, interference colors).

6.3.2.1 Color

Color is one of the most striking characteristics of the fine material. Care has to be taken, however, in its description. Color in PPL depends not only on the actual color of the fine material, but can vary also with the thickness of the section (see Section 3.1.3, Plate 3.1c and d)) and the temperature of the light used. The observed colors can be compared with those of a Munsell Soil Color Chart. More reliable determinations can be made with OIL since this technique is unaffected by the thickness of the thin section, although it is affected by the temperature of the light. As mentioned in Section 4.3.7 care should be taken when observing thin sections on a PC-screen, as colors strongly depends on the settings of the balance and the contrast.

Some substances are very strong pigments, and actually mask other minerals and their colors. The most common colors of fine material are discussed below:

Red: indicates the likely presence of fine dispersed hematite, but perhaps also some amorphous Fe-gels. Already small amounts (10%) of fine dispersed hematite result in a red color if the fine material (Baert and Van Ranst, 1997). Very large amounts of hematite, for example in laterites, may result in an opaque fine mass that is red in OIL and TDFI if finely dispersed (Table 3.2, Plate 3.2c and d). With decreasing thickness of the section, the red color in PPL tends to change to yellowish (Plate 3.1c and d). Heating, for instance by fire, of Fe-containing clay in oxidizing conditions results in a red color. Some organic substances may also have a reddish color (e.g., in excrements).

Yellowish brown: indicates the presence of finely dispersed goethite (Plate 3.2c), lepidocrocite or amorphous Fe gels and occasionally jarosite.

Brown to dark brown: mostly due to the presence of organic matter, commonly associated with Fe oxihydrates. Large amounts of humic colloids appear to be responsible for a sepia color.

Gray: the natural color of most clays and most fine-grained silicates and carbonates. The gray color reflects the absence of staining substances such as Fe and organic matter, for example in reducing environments.

Greenish to grayish green: indicates the presence of idiochromatic greenish minerals such as chlorite, nontronite, and glauconite (Plate

6.12a and b). A greenish staining by algal material has been observed in a few cases.

Black: localized, black (opaque) spots may be the result of impregnation by Mn oxihydrates or a concentration of finely dispersed ore minerals such as pyrite, graphite, or charcoal. The use of OIL is recommended (Plate 3.3d and f).

6.3.2.2. Limpidity

The transparency of the fine mass in PPL is called limpidity. The property is associated with the presence or absence of particles smaller than the thickness of the section that appear discrete because of opacity or relief. Complete absence of such particles results in a *limpid* aspect (gels and very fine clay) (Plate 6.12c and d); clay-size microcontrasted particles (mainly appearing as opaque specks) give a *speckled* aspect to the fine materials (Plate 6.2e, Plate 8.2a, Plate 8.14a, Plate 8.15a); fine silt-size opaque particles (e.g., strongly humified plant remains, charcoal or minute ore grains), give a *dotted* appearance (see also punctuations, Section 6.4.3). Some colorless or white very fine dispersed materials (e.g., gibbsite in Oxisols) may give a typical cloudy aspect to the fine material. Some materials (e.g., Mn-oxihydrates) cause opacity of the fine material.

6.3.2.3. Interference Colors

Interference colors of the fine material observed between XPL give some information about the mineralogical composition. Limpidity combined with isotropy generally indicates an amorphous fine material; high interference colors (white of higher orders) in a granular mass generally indicate the presence of fine carbonates (mainly calcite) (Plate 3.1e and f). When clay minerals have a strong parallel orientation (e.g., in clay coatings or a groundmass with striated fabric), their interference colors may reflect their composition: 1:1 clay minerals and hormites (sepiolite, palygoskite) show whitish or grayish interference colors of first-order (Plate 3.2a and b, Plate 8.6b), whereas pure 2:1 clay minerals have interference colors ranging from first order yellow to second-order colors (Plate 4.3a and b, Plate 6.12c and d, Plate 8.4b) (comparable to muscovite). Pure chlorite displays very low, mostly abnormal (blue or brown) interference colors. It should be noted, however, that these interference colors are often influenced by an admixture of amorphous materials, that reduces the total birefringence and by the color of staining material (e.g., yellowish or reddish staining by iron (hydr)oxides) giving rise to so-called "abnormal interference colors". When heated above their decomposition temperature, clays loose their birefringence, for instance in fire places and ceramics.

6.3.2.4. Other Characteristics

Several other specific characteristics can be useful in determining the nature of the fine material, or at least of some of its components. U.V.-fluorescence microscopy can be used to identify the presence of finely

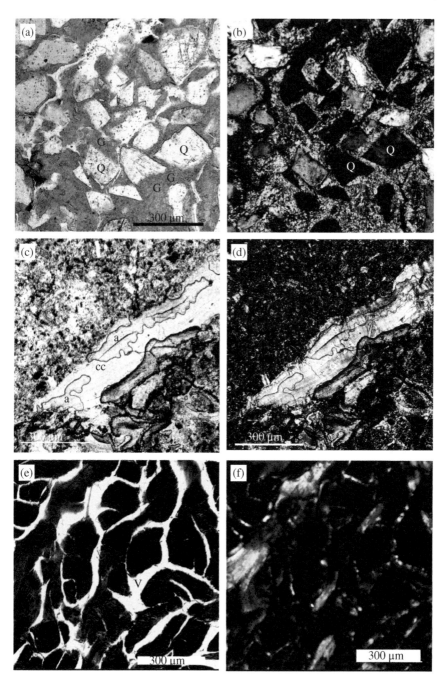

Plate 6.12. *Characteristics of fine material.* (**a**) groundmass of subangular, well sorted quartz grains (Q) in green, slightly speckled clay (glauconite) (G) with close porphyric c/f related distribution pattern (PPL); (**b**) same, notice high interference colors of the 2/1 clay and granostriated (gs) b-fabric (XPL); (**c**) entrapped fragment of coating (cc) of pure fine limpid smectitic clay; the irregular features with high relief are air inclusions (a) in the mounting medium (PPL); (**d**) same, notice perfect orientation of the fine clay and high interference colors (PPL); (**e**) fragment of old leather with shrinkage cracks (v) in Medieval archaeological deposits (PPL); (**f**) same, the gray interference colors in the voids are due to stress phenomena in the resin (XPL).

divided, more or less fresh, plant fragments (Babel, 1972), or the presence of Al-hydroxides (Van Vliet-Lanoë, 1980) (Plate 3.4c and d, Plate 6.13a and b). Cathodoluminescence can be used to distinguish different generations of carbonates (see Section 3.2.2.6, Plate 3.4e and f). Selective extraction, bleaching or staining of some components (Fe, calcite, organic matter) in uncovered thin sections may yield useful information with regard to the composition of the fine material (see also Section 3.2.3 and 3.2.4, Plate 3.5).

6.3.3. Size

Size determinations are only relevant for the coarser part of the fine fraction, depending on the quality of the available optical equipment and also on the contrast of the particles in relation to their surroundings: fine opaque particles can more easily be distinguished than transparent ones of the same size. Quality and thickness of the sections also influence the lower size limit of detection (see Section 3.1.4).

6.3.4. Shape

Particles in the fine material are generally too small to allow their shape to be observed by normal optical microscopy. Techniques such as scanning electron microscopy (SEM) and transmission electron microscopy (TEM) can be used, but a discussion of these methods is beyond the scope of this book. Only the shape of particles in the coarser part of the fine fraction can be described, and then only if the particles are sufficiently contrasted. Even under ideal conditions, description will generally be limited to estimating whether the particles are equidimensional or elongate (e.g., in the case of sericite flakes) (Plate 7.5). Observations in XPL may be helpful, as interference colors can enhance the contrast.

6.4. DESCRIPTION OF ORGANIC COMPONENTS
6.4.1. Introduction

Organic constituents are not only diagnostic components of the humus and of spodic horizons of soils, but often they also can be observed throughout the soil, as in the case of root remains or organic excrements. In general, only plant residues are observed, but sections through parts of animals have also been noticed in a few cases. Observation of bacteria in soil thin sections requires special pretreatments (e.g., Postma and Altemüller, 1990) and is beyond the scope of this manual.

The first morphoanalytical system to describe organic components was presented by Babel (1975) in a review paper. Attempts to develop a new terminology for organic matter, on the base of Brewer (1964a) were made by Bal (1973) and by Barratt (1969). Micromorphology of humus types was discussed already by Kubiëna (1948, 1953).

Compared with mineral components, organic components are much more difficult to describe because in most cases only morphology, color, and relief can be used as criteria, representing features that may change quite rapidly during humification.

6.4.2. Animal Residues

Very little information exists about animal residues in soil thin sections, except for their mineral parts (shells, bones, and skeletons), which were treated already in Section 6.2.4. Babel (1975) noted that the amounts of animal residues observed in soil thin sections are far below expectations.

Chitin fragments (shields of insects) are probably the most common items. Fresh fragments have distinct outlines and a pale yellowish color and relative high interference colors, and they can behave either as length slow or length fast particles; they are generally strongly fluorescent. In the soil, they are attacked by microorganisms and become dark brown to black (Babel, 1975). In archaeological materials, pieces of leather may be preserved under anaerobic conditions. They appear as a mosaic of brown to dark brown isotropic limpid substances (Plate 6.12e and f).

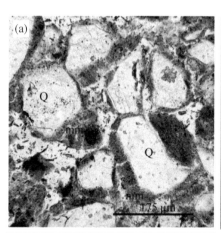

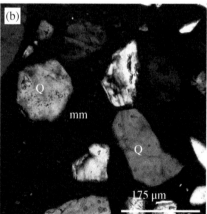

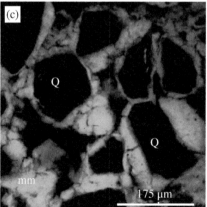

Plate 6.13. *Fine material* (a) coating of unoriented speckled clay (mm) on rounded quartz grains (Q); notice abrasive powder in voids (PPL); (b) same, notice undifferentiated b-fabric of micromass (mm) making a distinction between micromass and voids impossible (XPL); (c) same; the clay of the micromass (mm) is rich in short range order Al hydroxides and therefore shows strong fluorescence; quartz grains (Q) and voids remain black (BLF).

6.4.3. Plant Residues

The microscopic aspect of plant remains in soil thin sections depends largely on the drying conditions, as most contain high amounts of water (Kooistra, 2015). The oven and air drying result in shrinkage and darkening of the material, whereas acetone replacement in the liquid phase can result in the extraction of staining substances. The dark (up to opaque) color often hinders the exact recognition of humified plant remains.

The identification and description of plant remains in thin sections requires a sufficient knowledge of plant anatomy. Because this topic is, similar to the petrography of rocks, beyond the scope of this book, the reader is referred to several excellent handbooks and atlases on this topic, such as Bracegirdle & Miles (1971), Esau (1977), Schweingruber (1982), Gifford and Foster (1989), Schoch et al. (1988, 2004), Botanical Society of America (2008) and Upton et al. (2011).

According to Babel (1975) and Bullock et al. (1985), the organic constituents visible in soil thin sections can be subdivided into several types according to their size and complexity:

Organ residues: composed of at least five interconnected cells of more than one tissue type; sometimes the original contour of the organ is recognizable (e.g,. fragments of leafs, roots, stems, fruits and needles) (Plate 6.7a through d, Plate 6.14a through d, Plate 8.7e through g).

Tissue residues: composed of at least five interconnected cells of only one tissue type, and without a recognizable organ contour (Plate 6.14e, Plate 6.16c and d, Plate 8.18a). The most common tissue types according to Babel (1975) are:

> **Parenchymatic tissues**: composed of more or less equant thin-walled cells. The cellulose of the cell walls has first order gray interference colors (Plate 6.7a and b) that are commonly masked by the yellow color of the material. The brightness of the interference colors is not directly related to the degree of decomposition, as it also may vary in fresh materials. The cell lumen is generally empty except for plasma adhering to the cell walls. Fresh cell walls are strongly autofluorescent.

> **Lignified tissues**: composed of elongated, thick-walled originally empty cells. The cell lumen may contain colored alteration products.

> **Phlobaphene containing tissues**: with equant to oblate cells containing phlobaphenes in the cell lumen. These phlobaphenes show high chroma yellowish, brownish or reddish colors.

> **Plectenchyma** such as fungal pseudotissues, sklerotia (Plate 6.15a and b), and mycorrhiza mantles (Plate 6.15c and d).

Organic fine material: subdivided into:

> **Cells and cell residues**: other organic fragments of < five interconnected cells with recognizable, although quite often deformed, cells;

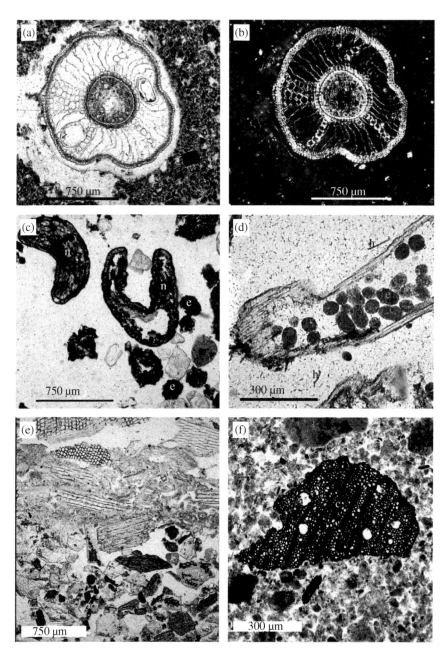

Plate 6.14. *Organ and tissue residues.* (**a**) section of fresh root (organ residue) (PPL); (**b**) same, notice strong interference colors of cell walls and undifferentiated b-fabric of the micromass (XPL); (**c**) needles (organ residues) (n) and organic fine material in spherical excrements (e) (PPL); (**d**) needle with fresh excrements (e) of organic fine material and fragments of hyphae (h) (PPL); (**e**) top: loose packing of tissue fragments; middle: tissue fragments in yellowish-brown amorphous fine material (am); bottom: coarse monic c/f related distribution of mineral grains and simple packing voids; cow sludge on sandy material (PPL); (**f**) fragment of charcoal (c) in groundmass with fine granular microstructure (PPL).

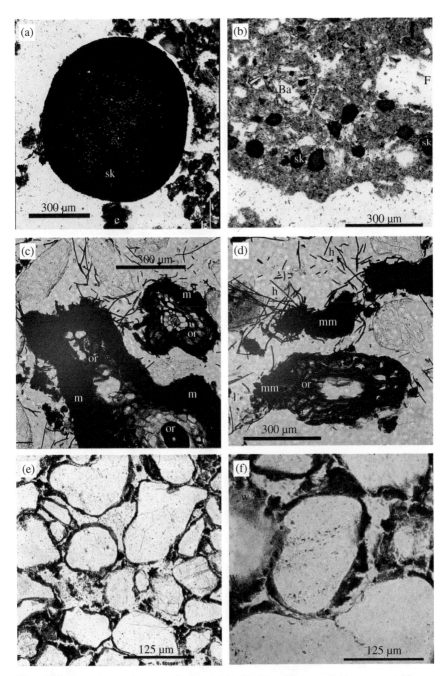

Plate 6.15. *Plectenchyma and monomorphic material.* (**a**) sklerotium (sk) surrounded by excrements (e) (PPL); (**b**) series of sklerotia (sk) with banded distribution pattern (PPL); (**c**) mycorrhiza mantle (m) on organic residue (or), and plenty of fungal hyphae (h) in packing voids (PPL); (**d**) mycorrhiza mantle (mm) on organic residue (or), and plenty of fungal hyphae (h) in packing voids (PPL) (**e**) chitonic c/f related distribution pattern: Rounded to subrounded quartz grains coated by amorphous organic material (monomorphic) (PPL) (**f**) same, detail; notice cracks (c) in organic coating due to drying and wedging (w) effect on grain at the left (PPL).

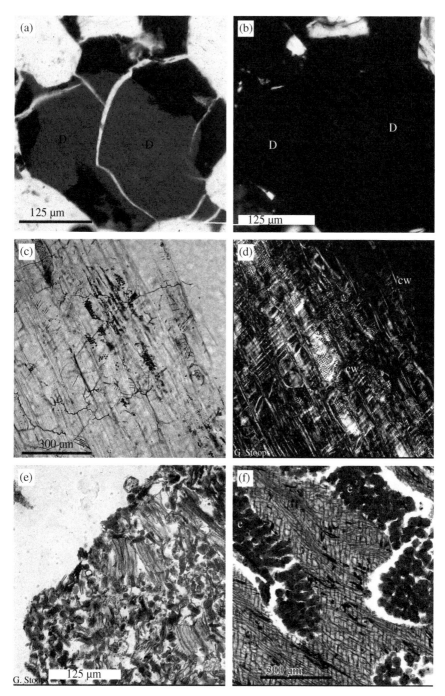

Plate 6.16. *Dopplerite and alteration of tissues.* (**a**) hyaline dopplerite (D) deposit; notice shrinkage cracks due to drying (PPL); (**b**) same; notice optical isotropism (XPL); (**c**) bleaching of tissue residue; notice presence of hyphae (h) (PPL); (**d**) same; notice interference colors of cell walls (cw) (XPL); (**e**) result of mashing of plants: Tissue (t) and cell (c) residues; part of excrement (PPL); (**f**) wood fragment partially eaten by mesofauna, from whom the fresh, purely amorphous organic excrements (e) are clearly visible (PPL).

this group also comprises fungal hyphae (Plate 5.2b, Plate 6.15c and d, Plate 8.18e and f) and their fragments, as well as spores and pollen grains.

Punctuations: small dark or opaque grains, about 1 μm large. They are part of the microcontrasted particles described in the fine material and contribute to a speckled or dotted limpidity of the fine material.

Organic pigment: occurs as stains in the fine material, generally brownish or grayish in PPL, darker than the surrounding material in OIL.

Amorphous organic fine material: can be of different types, such as mono- and polymorphicmaterial (see below). A typical example is the so called *dopplerite*, also called *waterhard* (Koopman, 1988; Ismail-Meyer, 2017) found beneath or around peat (Plate 6.16a and b). Organic excrements consist often of amorphous material (Plate 8.18b and c). With optical techniques it is not possible to distinguish between pure amorphous organic matter and organic colloids associated with colloidal Si, Al or Fe compounds, unless selective staining and/or extraction techniques are used.

Two main types of organic fine material have been described in Spodic horizons by De Coninck et al. (1973):

Monomorphic: amorphous organic fine material of uniform colloidal texture, with less than 5% inclusion of coarser organic elements (no recognizable vegetal or fungal structures), showing a system of polygonal desiccation cracks in sections of air- or oven-dried material, forming a continuous mass (Plate 6.15e and f);

Polymorphic: amorphous organic fine material with more than 5% coarser organic elements (cell residues, punctuations) included, but without recognizable vegetal or fungal structures; forming a discontinuous mass of manyform elements of different color and density. Note that the term polymorphic as used in this context is in no way related to the same term used in mineralogy.

6.4.4. Alteration

6.4.4.1 Introduction

There is no systematic scheme to describe the alteration of organic material. A system somewhat similar to the alteration classes proposed by Stoops et al. (1979) could be prepared in future for organic tissues. With respect to organ residues an effort was made by Blazejewski et al. (2005) to describe different stages of root decomposition in riparian soils, containing five classes, ranging from fresh roots to almost dispersed traces, taking into account breaks in the sheat, completeness of the inner portion, the amount of cellular structure still visible, the orientation of the remaining tissue fragments, and the degree of mixing with the groundmass. The effect of mesofauna (e.g., presence of excrements) and fungi are not considered. It would be helpful for thin section descriptions to have a similar system, but with extended alteration types (see Section 6.4.4.2) for other types of organs, such as leafs, stems or fruits.

As a result of wood fires or anthropogenic processes, plant fragments can be charred, preserving part of their fabric (Canti, 2017e; Brochier, 2016).

6.4.4.2 Alteration of Plant Tissues

According to Babel (1997) several types of alteration of organ- and tissue residues can be distinguished:

Deformation of cell structure: in still coherent residues, for example, as a result of shrinkage or chemical alteration.

Browning: due to formation of dye–protein complexes.

Bleaching: due to chemical decomposition of dyes, for example, by basidiomycetes (probably corresponding to white rot) (Plate 6.16 c and d).

Blacking: with preservation of tissue structure, mainly in water-logged conditions.

Gelifaction: the material is transformed to a gel-like material (probably corresponding to brown rot) (Plate 6.16.f).

Mashing: fragmentation by chewing soil-animals; the morphological aspects of tissues that have passed through the intestines of larger animals may remain unchanged, except for their sizes (Plate 5.2b, Plate 6.16e).

Dislocation of cells: due to the destruction of the intercellular material.

Colonization by fungi: replacing part of the plant tissue by fungal plectenchym.

Impregnation: for instance with iron hydroxides.

Disappearance: of part of the tissue due to mineralization or dissolution, visible as voids.

A systematic description of the size, color, interference colors, and internal structure of all components is necessary.

For more details on organic constituents, and their interpretation, the reader is referred to Stolt and Lindbo (2010), Kooistra (2015) Ismail-Meyer (2017), and Ismail-Meyer et al. (2018).

7. Groundmass

7.1 INTRODUCTION AND DEFINITION

Brewer and Sleeman (1960) were the first to divide the soil material into two distinct parts: the pedological features (e.g., clay coatings, iron nodules) and the enclosing material, which was called s-matrix in Brewer (1964a). Many of these features had already been recognized by other authors (e.g., Kubiëna, 1938), but Brewer (1964a) was the first to propose a systematic grouping, partly using some concepts from sedimentology. A similar way of thinking was followed by Bullock et al. (1985), but the terms groundmass and pedofeatures replaced s-matrix and pedological features respectively to emphasize the differences with Brewer's concept, and promote linguistically more correct terms.

> *Pedofeatures* are discrete fabric units present in soil materials recognizable from an adjacent material (the groundmass) by a difference in concentration in one or more components (e.g., a granulometric fraction, organic matter, crystals, chemical components) or by a difference in internal fabric.
>
> *Groundmass* is a general term used for the coarse and fine material and associated packing voids, which forms the base material of the soil in thin section, other than that in pedofeatures. (Bullock et al. (1985), modified)

According to these definitions the material left after identification of the pedofeatures (e.g., clay coatings, Fe-oxide nodules, calcite nodules and euhedral gypsum crystals) is the groundmass. The concept is not intended to apply to pedofeatures except in following cases: (i) where pedofeatures have arisen through impregnation or depletion of the groundmass which is still recognizable (matrix pedofeatures); (ii) in a detailed description of a pedofeature, where it is sometimes convenient to refer to the base material of the pedofeature as groundmass; (iii) when the pedofeature material is strongly dominant, for example, in the case of a gypsic or petrogypsic

Guidelines for Analysis and Description of Soil and Regolith Thin Sections, Second Edition. Georges Stoops.
© 2021 Soil Science Society of America, Inc. Published 2021 by John Wiley & Sons, Inc.
doi:10.2136/guidelinesforanalysis2

horizons consisting dominantly of gypsum crystals, this pedogenic material can be considered as groundmass to identify other pedofeatures (e.g., nests of celestite crystals).

Although the attention of students is automatically drawn to pedofeatures, a full analysis of the groundmass is most important. Indeed, whereas pedofeatures reflect past and present pedogenic processes, the groundmass reflects in addition the lithology, the homogeneity and the degree of weathering of the parent material (Stoops and Mees, 2018). Moreover, some soil materials (e.g., in young soils and in Oxisols) do not contain pedofeatures; the groundmass is then the only source of information.

> BACKGROUND – Brewer (1964a) recognized an *s-matrix* composed of skeleton grains, plasma and associated voids. He defines *skeleton grains* as "individual, relatively stable grains that are not readily translocated, concentrated or reorganized by soil forming processes". Complex grains, such as rock fragments, were not considered as skeleton grains, but as *pedological features*. *Plasma* was "that part of a soil material that is capable of being moved or having been moved, reorganized and/or concentrated by soil forming processes". This includes all the materials, mineral or organic, of colloidal-size (i.e., < 2 µm) and the relatively soluble materials not bound up in skeleton grains. Thus, plasma includes colloid-size grains, solubles such as carbonates, and fractions of the humified organic matter. The difference between plasma and skeleton grains was thus based on absolute size (with a 2-µm limit) and stability or solubility. The latter created problems, especially in the case of arid soils, where, for example minerals such as calcite, or even gypsum may be stable under arid conditions, but are most unstable in the tropics. In his later publications the author became aware of this problem, and used the concept of framework members (*f-members and f-matrix*) (Brewer and Pawluk, 1975, Brewer and Sleeman, 1988). A practical difficulty is the size limit of 2 µm, which is below the resolution for noncontrasting particles in thin sections. As a matter of fact, in literature most descriptions of plasma deal with grain sizes up to 5 or 10 µm. For a more detailed discussion of these topics, see Stoops and Jongerius (1975).

The difference between s-matrix and groundmass and between pedological features and pedofeatures is caused by Brewer's subdivision of the soil material into skeleton grains and plasma. Rock fragments, weathered grains and fragments of inherited nodules are excluded from the s-matrix as defined by Brewer (1964a), whereas they are part of the groundmass as defined by Bullock et al. (1985). Soluble grains are part of the plasma in Brewer's system, whereas they can belong to the coarse fraction in Bullock et al. (1985).

7.2 DESCRIPTION

The description of the groundmass includes the following items: (i) the limit between coarse and fine material, (ii) the c/f related distribution, as far as it was not yet described as a microstructure, (iii) the coarse material and its fabric, and (iv) the fine material (micromass) and its fabric. Packing voids are dealt with in Chapter 5.

7.2.1 The Limit Between Coarse and Fine

While the simplest soil material (e.g., dune sand) consists of only two types of fabric units: (mineral) grains of one size class and associated voids, in most soils different grain sizes (e.g., sand and silt and/or clay) are present. It seems logical to use grain size as one of the criteria for distinguishing fabric units, as is done for instance in petrography (e.g., phenocrysts and matrix). As very fine particles (e.g., clay, amorphous material) have specific physical, chemical and mechanical characteristics (e.g., cation exchange capacity, possibility to be translocated in the profile), absent or less apparent in coarser size fraction (e.g., sand), it is evident that soil scientists attribute a special importance to them.

In view of the enormous differences among soil materials, it is preferable not to have a standard size limit between coarse and fine material. In some highly weathered soils, 2 μm is a possible limit, whereas in most (semi)arid soils 5 or even 10 μm would be preferable, to include the fine calcite grains (5- 10 μm) in the fine fraction. Also in A horizons, with many organic fine particles, a 2-μm limit is not realistic. Moreover, when studying thin sections at low magnification, the recognizable size limit between coarse and fine material will be automatically higher than 2 μm, because it will be beyond the resolving power of the optical microscope (see Section 3.1.4). It is not logical to select a size limit that cannot be applied with the technique employed.

For a given thin section, or set of thin sections, the size limit between coarse and fine materials can be chosen taking into account following criteria:

1. Resolving power of the microscope at the highest magnification to be used.

2. Size distribution of the material present (e.g., in a soil material consisting of sand and silt only, it is useless to put the limit at 2 or even 5 μm).

3. Kind of material present; commonly the fine and coarse materials are each characterized by a different mineralogical nature.

4. Size limits used in the laboratory (I.U.S.S.-system, USDA-system, DIN or others).

5. The objective of the study.

The size limit between coarse and fine can be indicated by adding a subscript to the letter group c/f. For example $c/f_{10\mu m}$ indicates that the limit between coarser and finer material is 10 μm (Stoops and Jongerius, 1975). The general symbol for the *c/f limit* is thus c/f_x. A very useful indication is the *c/f-ratio*, expressing the ratio between the volume occupied by the coarse material and that by the fine material. For example, a $c/f_{2\mu m}$ ratio of 2/3 indicates that for two volumes of material coarser than 2 μm, three volume units of finer material are present. More precise data can be obtained by micromorphometric methods (e.g., point counting or image analysis). More than one c/f-ratio can be given subsequently if the groundmass contains particles belonging to clearly separated size classes. For example, in a soil on gravel the c/f_{5mm} limit distinguishes between grains coarser and finer than 5mm (e.g., between coarse sand or gravel and the interstitial material) whereas a $c/f_{5\mu m}$ limit can be used to describe the interstitial material. The fine material of the groundmass forms a partial fabric, which is called the *micromass* (Karale et al., 1974).

7.2.2 The c/f related Distribution Pattern

The c/f related distribution, as described in Section 4.3.2.4, is one of the most important characteristics of the groundmass. The concept has been frequently applied with success in pedogenic research (e.g., Chadwick and Nettleton, 1994). Note that most c/f related distribution patterns, except the porphyric, are described also as microstructures. For example, a chitonic c/f related distribution pattern, consisting of sand grains surrounded by clay coatings, is described as a pellicular grain microstructure. The groundmass in this case comprises only the sand grains, whereas the clay coatings are pedofeatures. Thus the real c/f related distribution in the groundmass is actually coarse monic.

In Table 7.1 the c/f related distribution in the groundmass is compared with similar concepts in other systems.

7.2.3 Coarse Material, Composition and Fabric

For coarse material it is essential to describe the composition (see Section 6.2), size, and shape of every mineral type present. For

Table 7.1. Related distributions (r.d.) of coarse and fine material in the groundmass according to different systems.

Coarse and fine constituents (relative sizes)		plasma and skeleton grains (size limit at about 2 µm)		
Stoops & Jongerius (1975).	Brewer & Sleeman (1988)	Kubiëna (1938)	Brewer (1964a)	Eswaran and Baños (1976)
c/f r.d	r.d. of f-members and f-matrix	elementary fabric of plasma and skeleton	r.d. of plasma and skeleton grains	NRDP and SRDP†
monic	granic	bleached sand (partially)	granular	granic
chitonic	chlamydic	chlamydomorphic	-	dermatic
gefuric	iunctic	intertextic, plectoamictic (partially).	intertextic (partially).	intertextic
enaulic	enaulic agglomeratic	agglomeratic	agglomeroplasmic	congelic
porphyric	porphyric	porphyropeptic and porphyropectic	*porphyritic (1960)* porphyroskelic (1964a)	porphyric, plasmic

† NRDP: Normal Related Distribution Pattern; SRDP: Special Related Distribution Pattern.

example, rounded quartz grains belonging to the coarse sand fraction, and subangular feldspar of medium to fine sand size. In most cases, the arrangement is random. If not, the orientation and distribution patterns should be indicated (e.g., banded in the deeper parts of Fluvisols, banded with graded bedding in glacial deposits and sedimentary surface crusts).

The distinction between nodules, considered as pedofeatures, and nodular bodies, considered to be of detrital origin, and thus described as coarse material, depends largely on the experience of the scientist. A few guidelines are given in Section 8.6.2. For more details and interpretation of the coarse components see Stoops (2015b) and Stoops and Mees (2018).

7.2.4 The Fine Material (Micromass) and Its Fabric

7.2.4.1. Introduction and Definition

Because the nature of the micromass generally cannot be determined by its optical characteristics, its description includes the following: color, limpidity, and if present, its interference colors (see Section 6.3). More information can be gained using special techniques, such as UVF, selective chemical extractions (see Chapter 3) and micro-analyses such as WDS, micro-XRD, micro-FTIR.

The micromass consists generally of crystalline and/or short range order clay minerals, often combined with Fe-(hydr)oxides, amorphous and/or fine organic matter, and in some cases with crystallites of calcite or sericite flakes. Because the former materials are beyond the resolution of the optical microscope at the scale of observation, their fabric elements (e.g., size and shape of the particles, and their orientation and distribution patterns) cannot be observed directly. In Section 4.3.2.2.3. it was mentioned that zones of clay particles with a parallel basic orientation pattern display interference colors (orientation birefringence) when observed in thin sections between crossed polarizers. Such a parallel orientation is a rule rather than an exception in natural clays. Clay particles form small aggregates (20–30 μm) with an internal parallel basic orientation pattern. The aggregates were called domains by Aylmore and Quirk (1959) and Smart (1969), and pseudocrystals by Soviet micromorphologists (Dobrovol'ski, 1983), as explained in Section 4.3.2.2.3, and in Stoops and Mees (2018).

By focusing on clay domains, an indirect description of the fabric of the fine material is possible, if we look for the absence or presence of interference colors in the micromass and their pattern. As interference colors are the result of birefringence of the particles, the term birefringence fabric or *b*-fabric was introduced by Bullock et al. (1985).

> *The b-fabric* is the patternof orientation and distribution of interference colors in the micromass (Bullock et al., 1985, modified).

BACKGROUND - The concept of *b*-fabric is related to the plasmic fabric of Brewer (1964a and b), the birefringent streaks

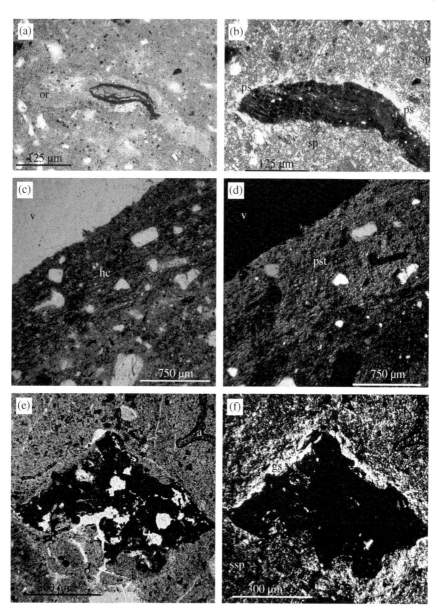

Plate 7.1 *Striated b-fabrics* (**a**) organ residue (or) (root) in yellowish speckled groundmass; different types of individual cells are clearly recognizable (PPL); (**b**) same, notice weak interference colors in the root and porostriated b-fabric around the root channel (ps) and stipple speckled (sp) *b*-fabric in the rest of the groundmass (XPL); (**c**) hypo-coating (hc) of organic fine material oriented parallel to the void (v) surface (PPL); (**d**) same, a striated orientation of clay domains parallel to the void (v) surface is responsible for porostriated *b*-fabric (pst); stipple speckled *b*-fabric in the lower right corner (XPL); (**e**) opaque grain in brownish groundmass; features with high relief are air bubbles in the mounting medium (a) (PPL); (**f**) same, granostriated *b*-fabric (gs) around the grain and stipple speckled (ssp) *b*-fabric in the rest of the groundmass (XPL).

(Doppelbrechende Schlieren) of Kubiëna (1938), the microtexture of some Russian sedimentologists, the microstructure of Russian soil micromorphologists and the (micro)fabric and microstructure of soil engineers. FitzPatric (1989) opposed the use of the term *b*-fabric as birefringence is a numerical value and not a visible feature. He uses the term anisotropic to indicate the presence of zones with interference colors in the fine mass, but he does not propose a specific subdivision. As the term *b*-fabric is already in use, changing its name would lead to confusion. Moreover, the term birefringence has been used in a nearly similar sense in earth sciences in several languages.

The *b*-fabrics are excluded from the pedofeatures although few of them (within the striated types) could be included according to the definition of (fabric) pedofeatures.

7.2.4.2. Types of *b*-fabric

Five main types are distinguished: undifferentiated, crystallitic, speckled, striated, and strial.

a. **Undifferentiated *b*-fabric:** characterized by an absence of interference colors in the fine mass (Plate 6.1b, Plate 6.13b, Plate 6.14b, Plate 8.13b and d). This absence can be caused by: (i) the dominance of short-range order clay minerals (e.g., allophane in volcanic ash soils), (ii) the masking by humus (e.g., in many A horizons) or sesquioxides (e.g., in Oxic materials), (iii) the presence of opaque material (e.g., Mn-oxides), or (iv) the random orientation of the clay particles causing a statistical isotropy.

b. **Crystallitic *b*-fabric:** characterized by the presence of small birefringent mineral grains (e.g., calcite or sericite) that determine the interference colors of the fine mass as a whole (Plate 3.1e and f; Plate 6.10b, Plate 6.11e and f, Plate 8.1f, Plate 8.9f, Plate 8.19b, Plate 8.21b through f). Specific orientation and distribution patterns (Plate 7.5a and b) are described as for striated fabrics (see below). It is also necessary to mention the nature of the minerals creating this fabric (e.g., calcitic crystallitic, sericitic crystallitic or gibbsitic crystallitic *b*-fabric).

c. **Speckled striated, and strial *b*-fabrics** When the fine mass is dominantly composed of anisotropic clay, zones of birefringence produced by particle orientation are frequently observed in XPL by their interference colors. If these zones consist of randomly oriented small (a few micrometers), equant domains, the term *speckled b-fabric* is used; if they consist of small elongated domains, called striae, with a parallel orientation, forming streaks, sometimes several hundred of micrometers long, the term *striated b-fabric* is used. A general orientation of the clay material usually inherited from sedimentation processes is referred to as a *strial b-fabric*. The orientation of the striae can be determined easily with the help of a retardation plate (gypsum sensitive plate), given that phyllosilicates are always length slow (Section 4.3.2.2.3).

c.1 Speckled b-fabric: characterized by randomly arranged, equidimensional or slightly prolate domains of oriented clay. When the microscope stage is rotated, individual domains extinguish successively, whereas other appear, so that the general aspect of the fine mass is not changed. Two subtypes are recognized:

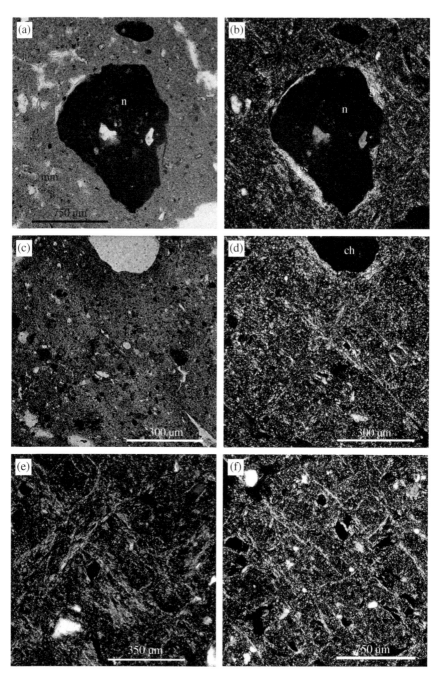

Plate 7.2. *Striated b-fabrics* (**a**) anorthic (lithomorphic) iron (hydr)oxide nodule (n) in grayish brown micromass (mm) (PPL); (**b**) same, well expressed granostriated (gs) *b*-fabric around the nodule (n), and stipple speckled in the groundmass (XPL); (**c**) brown dotted micromass with some short planes (p) and channels (ch) (PPL); (**d**) same, notice porostriated (around channel), granostriated (around nodule, top), monostriated (ms) and stipple speckled *b*-fabric (XPL); (**e**) cross-striated *b*-fabric oriented at about 45° (XPL); (**f**) cross- and granostriated (gs) *b*-fabric; the cross-striated *b*-fabric is best expressed in the NW-SE direction, making an angle of 45° with the analyzer and polarizer (XPL).

c.1.1. Stipple-speckled: consisting of individual and isolated speckles (Plate 6.13a, Plate 7.1b and d, Plate 7.2d, Plate 8.1b and f),

c.1.2. Mosaic-speckled: the speckles are in contact with each other, resulting in a mosaic-like pattern (comparable to the internal fabric of some glauconite grains).

c.2. Striated b-fabrics: characterized by the presence of elongated zones or streaks, at least 30 μm long, in which the domains show more or less simultaneous extinction. The streaks do not show continuous interference colors, but are built up by a juxtaposition of smaller, more or less parallel arranged prolate domains of oriented clay particles (striae). The streaks therefore lack sharp boundaries and normally cannot be distinguished in PPL, as this would indicate an accumulation of clay (pedofeature), rather than a pure orientation feature. Striated patterns are common in clayey soils subject to shrink and swell (Dalrymple and Jim, 1984; Jim, 1990), or freezing (Huijzer, 1993). They are the result either of shearing or of confined pressure, as demonstrated in soil-mechanic studies by Morgenstern and Tchalenko (1967b) and Mitchell (1956) (for more details see Stoops and Mees, 2018). They generally only develop when the $c/f_{5\mu m}$ related distribution type is situated below the surface of limit of skeletal function (see Fig. 4.8). Between the striated patterns, a speckled *b*-fabric is present.

Striated *b*-fabrics are subdivided mainly on the basis of the orientation and distribution patterns of the streaks (Fig. 7.1).

c.2.1. Porostriated *b*-fabric: clay domains in the micromass are oriented parallel to the surface of a void, as in the case of slickensides (Plate 7.1a through d; Plate 7.2c and d).

c.2.2. Granostriated *b*-fabric: clay domains are oriented parallel to the walls of resistant fabric units (mineral grains, hard nodules) producing a halo of interference colors around the grains in XPL (Plate 3.5b and e, Plate 7.1e and f; Plate 7.2a and b, Plate 8.1b, Plate 8.14d, Plate 8.15b and d). In some cases two tails oriented in oppo-

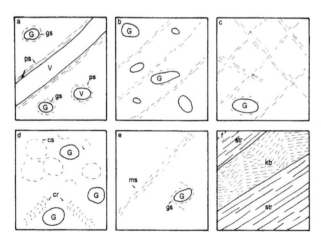

Fig. 7.1. Striated and strial *b*-fabrics: (**a**) poro- and granostriated (ps and gs), (**b**) parallel striated, (**c**) cross-striated, (**d**) circular striated (cs, top) and crescent striated (cr, bottom), (**e**) monostriated (ms) and granostriated with galaxy structure, (**f**) unistrial (str) with kinking b-fabric (kb). G, grain; V, void.

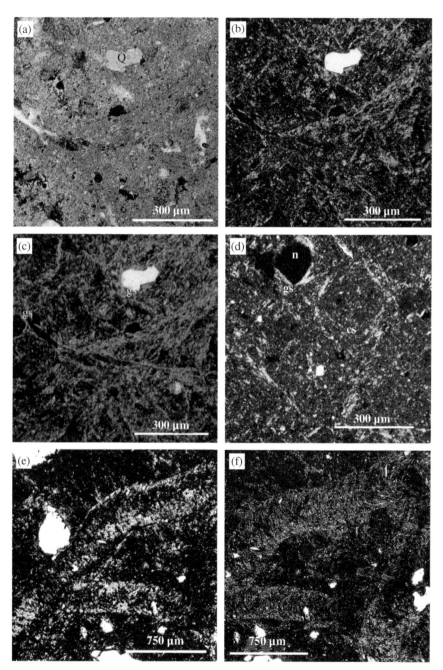

Plate 7.3 *Striated b-fabrics* (**a**) yellowish speckled micromass and angular quartz (Q) grains, forming an open porphyric c/f related distribution pattern (PPL); (**b**) same, striations occur according to all directions, yielding a random striated *b*-fabric (XPL); (**c**) same, the manifold orientations of the striations becomes more pronounced in CPL; (**d**) granostriated *b*-fabric (gs) around Fe-hydroxide nodules (n) and circular striated (cs) *b*-fabric; the expression of the *b*-fabric is exaggerated by a long exposure time, which also caused the grayish background in the voids (XPL); (**e**) crescent striated *b*-fabric ; open porphyric c/f related distribution (XPL); (**f**) same, notice absence of extinction in CPL.

site directions formed as a result of stress-induced rotational movements of coarse grains can be observed (Lafeber, 1964). The latter type is closely related to "galaxy structures" (van der Meer, 1997) (Section 4.3.2.4.2). Porostriated and granostriated *b*-fabrics are also related to fabric hypo-coatings (see Section 8.2.2), but not considered as pedofeatures.

 c.2.3. Monostriated *b*-fabric: isolated and independent streaks of parallel oriented clay domains are observed in the micromass.

 c.2.4. Parallel striated *b*-fabric: the streaks occur in parallel or subparallel sets, where they show a more or less simultaneous extinction when the microscope stage is rotated (Plate 7.2a and b).

 c.2.5. Cross-striated *b*-fabric: two sets of streaks intersect (Plate 7.2e and f).

BACKGROUND - If the sets of streaks intersect at right angles, the term reticulate striated b-fabric was proposed by Bullock et al. (1985). As the angle of intersection depends on the absolute orientation of the section, the use of the latter term is discouraged as it has no objective value.

 c.2.6. Random striated *b*-fabric: an irregular pattern of interwoven fine streaks, which extinguish in different positions when the microscope stage is rotated (Plate 7.3a through c).

 c.2.7. Circular striated *b*-fabric: streaks occur as more or less circular features or rings (e.g., in some oxic horizons) (Plate 7.3d).

 c.2.8. Concentric striated *b*-fabric: streaks occur as concentric rings (e.g., in the Leda clay, Ontario).

 c.2.9. Crescent striated *b*-fabric: the streaks represent elongated, worm-like features with a crescent-like (bow-like) internal fabric (common in ferrallitic soils) (Plate 7.3e and f). A transversal section through such worm-like feature yields a local concentric striated *b*-fabric.

To complete a general description of striated *b*-fabric the following additional information is desirable:

Continuity of the birefringent streaks: continuous or discontinuous,

Thickness: thin (< 20 μm), medium (20–200 μm) and thick (> 200 μm),

Length: (except for features related to surfaces and for circular and concentric striated fabrics).

More detailed descriptions of the striated *b*-fabric involve the following determinations:

Minimum size of the coarse particle surrounded by granostriated b-fabric: development of this *b*-fabric is generally function of the size of the coarse grains (Blokhuis et al., 1970).

Degree of orientation: three degrees of striated orientation are distinguished according to the percentage occupied by parallel oriented domains: strongly striated (> 66%), moderately striated (between 66 and 33%) and weakly striated (< 33%) (see Section 4.3.2.2.2).

Frequency: expressed as a percentage of the total surface of the fine mass. Two measurements are possible:

 (i) whereby the percentage is determined for a fixed position of the rotating stage (i.e., only the streaks not parallel to the polarizer or to the analyzer are taken into account),

(ii) the percentage is determined with circular polarized light, or by rotating the stage, enabling all streaks, independently of their orientation, to be observed (e.g., Embrechts and Stoops, 1986).

Absolute orientation: with reference to the soil surface for example. This determination is particularly useful in the case of mono- parallel and cross-striated types, although it is function of the orientation of the thin section (see Section 3.1.6) (Magaldi, 1974). Proposals for a graphical presentation of the orientation patterns are given by Hill (1970).

c.3. Strial b-fabric: (compare strial plasmic fabric, Brewer 1964a and b): the micromass exhibits as a whole a preferred parallel orientation (observed in unconsolidated sediments, deeper horizons of clayey soils and sometimes also in depositional crusts). The overall absolute orientation should be examined with the sensitive retardation plate (Plate 7.4e). Two subtypes are distinguished:

c.3.1. **Unistrial:** one preferred direction (Plate 7.4a trough e).

c.3.2. **Bistrial:** two intersecting preferred directions.

When the section is rotated until the strial fabric is in extinction, a superposed striated or speckled pattern may become visible, which has to be described separately. An indication of the absolute orientation (e.g., with reference to the soil surface) may be important for interpretation.

c.3.3. **Kinking fabric:** Bordonau and van der Meer (1994) observed in glaciolacustrine deposits a fabric superposed to unistrial *b*-fabrics, characterized by parallel, alternating bands more or less perpendicular to the bedding planes, approximately 300 to 700 µm thick. The fabric is the result of a herringbone arrangement of domains of clay particles due to crenulation cleavage (Plate 7.4f).

BACKGROUND - Compared to Bullock et al. (1985) the following alterations have been made: a minimum size of 30 µm for domains was introduced to distinguish striated from speckled *b*-fabrics; reticulate striated *b*-fabrics were deleted and incorporated in the cross-striated type, as the angle of intersection depends on the relative orientation of the section, and a kinking *b*-fabric was introduced. Brewer (1964a, 1964b) distinguished a series of *plasmic fabrics*, which describe the arrangement of the plasma (see Section 7.1) in the s-matrix., on the basis of the pattern of extinction figures between crossed polarizers. Plasmic fabrics are therefore comparable to the *b*-fabrics of Bullock et al. (1985), but they are not synonymous because "plasma" is not equal to micromass with regard to size and composition. Russian sedimentologists and micromorphologists also recognized several microfabrics. Table 7.2 compares these different terminologies.

7.2.4.3. Factors Influencing the Expression of *b*-Fabrics

When using only XPL, observations should be made while rotating the stage. In circular polarized light (CPL) some striated types may be better expressed (e.g., circular striated or crescent striated), others may merge to a global random striated *b*-fabric. (Plate 7.3c and f).

Thickness of the thin section may play an important role in the determination of crystallitic fabrics. Thick sections suggest a close packing, even overlapping of crystallites, whereas thinner sections (e.g., 10 µm) can

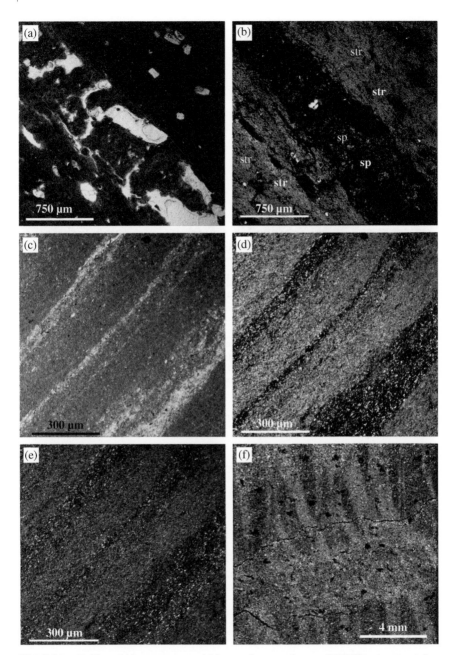

Plate 7.4 *Strial b-fabrics.* (**a**) homogeneous grayish brown micromass; the central NW-SE zone corresponds
to an infilling by excrements (PPL); (**b**) same, notice the strial *b*-fabric (str) (oriented NW-SE) of the
groundmass and the weakly stipple speckled *b*-fabric (sp) of the excrements in the central zones with identical
composition (XPL); (**c**) layered sediment (wharves) with banded distribution pattern of the coarse and fine
components; the section was turned over 45° to obtain a maximum visibility of interference colors (PPL);
(**d**) same, strial *b*-fabric with SW-NE orientation in the micromass (XPL); (**e**) same, but with retardation
plate; all domains show higher (blue) interference colors demonstrating that they are all oriented in the
same direction (SW-NE) (XPLλ, nγ oriented SW-NE); (**f**) unistrial *b*-fabric showing well developed kinking
b-fabric as a result of shears, visible as an alternation of subvertical darker and lighter streaks; Weichselian
glaciolacustrine rhythmites (XPL).

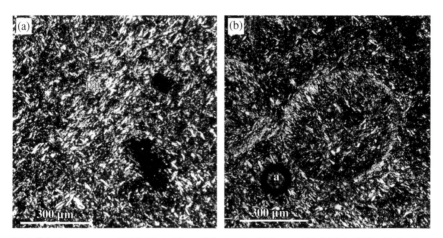

Plate 7.5 *Micaceous crystallitic b-fabrics* (**a**) parallel striated sericitic crystallitic *b*-fabric (XPL); (**b**) circular striated sericitic crystallitic *b*-fabric; air bubble in mounting medium (a) (XPL).

Table 7.2. *b*-fabrics of the micromass according to different systems.

Bullock et al. (1985) *b*-fabric	Brewer (1964a and b) plasmic fabric	Dobrovol'ski (1983) microtexture
undifferentiated	isotic	-
crystallitic	not recognized; partially silasepic with calcite; the term crystic is often used, but not correct in this context	-
speckled	asepic	scaly
stipple speckled	argilasepic and insepic	-
mosaic speckled	mosepic	± scaly-fibrous
striated	sepic	fibrous
porostriated	vosepic	-
granostriated	skelsepic p.p.	-
monostriated	-	-
parallel striated	masepic	parallel fibrous, flow fibrous
crossstriated	clino-bimasepic	cross fibrous
(reticulate striated)	bimasepic, partially lattisepic	-
random striated	omnisepic	random fibrous
circular striated	-	-
concentric striated	-	concentric fibrous
crescent striated	-	-
strial	strial	-
-	crystic (for pure crystalline features)	-
-	undulic (with a general undulose extinction)	-

show a relative large spacing between the crystallites, and a transition to a speckled or striated pattern appears (Plate 3.1e and f). In the case of crystallitic *b*-fabrics caused by the presence of calcite crystallites, removal of the carbonates (see Section 3.2.3.2) may reveal a masked striated or speckled fabric (e.g., appearance of striated *b*-fabrics in the case of calcic Vertisols, Plate 8.1 e and f).

The visibility of striated and/or flecked patterns depends on the thickness of the thin section, the intensity of the light source (e.g., use of substage condenser) and the magnification. Increasing thickness of the section and decreasing light intensity will reduce the apparent degree of development of the fabric, which will change from striated to speckled and finally to undifferentiated. Wedging effects (see Section 3.1.3) may suggest porostriated and especially granostriated *b*-fabrics as a result of a reduction of the thickness of the clay layer on oblique surfaces. This pitfall can be avoided by making use of the sensitive retardation plate, or comparing with the fabric around opaque grains.

Undifferentiated *b*-fabrics may become striated or flecked ones if iron or organic compounds are removed (see Section 3.2.3), if thickness of the section is reduced or if light intensity is increased. In the same way stripping the Fe (hydr)oxides and/or organic colloids may enhance striated and speckled *b*-fabrics (Plate 3.5).

Several types of striated *b*-fabrics, especially parallel and cross-striated, can be an optical illusion, induced by the extinction of streaks parallel to the polarizer or analyzer and a maximum interference of those inclined at 45°. When circular polarized light (CPL) is used, the section tends to show a random striated *b*-fabric in these cases. This does not preclude, however, the existence of real parallel and cross-striated fabrics, for example in vertic soil materials.

Striated fabrics may also originate by the shearing of former clay coatings. Provided the presence of clay accumulation can be assumed (e.g., based on observations in PPL), these streaks are considered as deformed pedofeatures (see Section 8.11). In clayey soils well-developed striated *b*-fabrics may be formed artificially at the spot where wet samples have been sheared by a spade, knife or Kubiëna box.

It is also important to describe the degree of expression, e.g., "very weakly expressed" means that high magnification and strong transmitted light are needed to make it visible, whereas "strongly expressed" indicates that the fabric is already visible at lower magnification and light intensity. In describing the development of speckled, striated and strial *b*-fabrics, a clear distinction should be made between strong interference colors, pointing to a strong orientation of anisotropic clay particles, and high interference colors, related to the nature of the clay (see also Section 6.3.2.3).

The relationship between flecked and striated *b*-fabrics, and the mechanical properties of the soil material has already been shown for instance by Cagauan and Uehara (1965), Embrechts and Stoops (1986), and Zainol and Stoops (1986). More detailed studies on this topic are needed.

8. Pedofeatures

8.1 INTRODUCTION AND DEFINITIONS

As mentioned in Section 7.1, Brewer and Sleeman (1960) introduced the concept of pedological features to indicate specific organizations in the soil material, as opposed to the undifferentiated s-matrix. Bullock et al. (1985) adopted this point of view but added new subdivisions and adapted others. Brewer's term pedological features (literally "soil science features") has been replaced by pedofeatures, for linguistic reasons and to emphasize the basic differences between both systems.

> *Pedofeatures* are discrete fabric units present in soil materials that are recognizable from groundmass by a difference in concentration in one or more components or by a difference in internal fabric. (Bullock et al., 1985)

Examples of pedofeatures are clay coatings, calcite nodules, euhedral gypsum crystals, excrements of the soil fauna, goethite mottles, and passage features. The b-fabrics are not considered as pedofeatures, although some striated types, could fit the criteria.

There is no upper size limit for pedofeatures, although if the feature is continuous throughout a thin section (e.g., placic horizon), or covers the whole of the section (e.g., lateritic crust, petrocalcic horizon), it is more reasonable to treat it as a horizon or layer and describe it from this standpoint. The lower size limit is about 20 μm; below this the features cannot be recognized readily in thin sections with an optical microscope.

BACKGROUND – The difference between Brewer's pedological features and Bullock's pedofeatures is based on the definition of skeleton grains (Brewer and Sleeman, 1960) as simple individual grains (see also Section 7.1). Rock fragments, weathered grains, and fragments of nodules from eroded soils are therefore excluded from the s-matrix and have to be considered as pedological features. In Bullock et al. (1985), these elements are considered as part of the coarse fraction of the groundmass. In Russian research, the term neoformation is used for pedofeatures.

Guidelines for Analysis and Description of Soil and Regolith Thin Sections, Second Edition. Georges Stoops.
© 2021 Soil Science Society of America, Inc. Published 2021 by John Wiley & Sons, Inc.
doi:10.2136/guidelinesforanalysis2

8.2 SUBDIVISION OF PEDOFEATURES

8.2.1 Introduction

Based on Stoops (1998) a distinction is made between pedofeatures result-ing from a change of the groundmass, called matrix pedofeatures, and those formed outside the groundmass, called intrusive pedofeatures. On another, equivalent level, pedofeatures are subdivided according to their morphol-ogy. Only a combination of both classifications gives a satisfactory identifi-cation of the features.

8.2.2 Matrix and Intrusive Pedofeatures

Matrix pedofeatures are pedofeatures that result from a change in com-position or fabric of the groundmass; this implies that several characteristics of the groundmass remain visible in the pedofeature. Most commonly the nature and distribution pattern of the coarse fraction is preserved (Stoops, 1998, modified.).

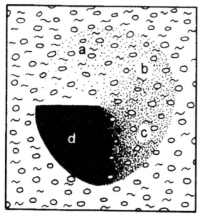

Fig. 8.1. Degree of impregnation of matrix pedofea-tures: weakly impregnated (characteristics of the origi-nal groundmass are dominant) (**a**); moderately im-pregnated (characteristics of the original groundmass are common and clearly identifiable) (**b**); and strongly impregnated (the material of the pedofeature is domi-nant. but the nature of the groundmass (e.g., quartz grains) can still be identified); (**c**) there is no relation between the feature and the groundmass (**d**).

Three subtypes of matrix pedofea-tures can be distinguished (Table 8.1):

Impregnative pedofeatures: recog-nizable because of a higher concen-tration of a component, generally in the micromass, such as amorphous Fe- or Mn-(hydr)oxides, fine-grained calcite or amorphous organic matter. However, the use of the term impreg-native does not imply any interpreta-tion of the process of their formation. These pedofeatures may be the re-sult of an impregnation of the micro-mass (e.g., by colloidal Fe hydroxides) but they can also be the result of an alteration (e.g., a local oxidation of Fe-compounds), an epigenetic replace-ment with preservation of the volume of the micromass (Turc et al., 1985), or an expulsion of the micromass (e.g., clay by calcite). Impregnative pedofea-tures generally require microporosity

Table 8.1. Key to the classification of pedofeatures according to their relation with the groundmass (Stoops, 1998).

1. Pedofeature still presenting the general characteristics of the groundmass, mostly, but not always, with a gradual transition to the groundmass:	matrix pedofeatures
1.1 Matrix pedofeatures recognizable because of a lower concentration of one or more components:	depletion pedofeatures
1.2 Other matrix pedofeatures recognizable because of a higher concentration of one or more components:	impregnative pedofeatures
1.3 Other matrix pedofeatures recognizable because of a different fabric (with exclusion of bfabrics):	fabric pedofeatures
2. Pedofeature non-enclosing the groundmass, formed either in voids or in the groundmass by accretion:	intrusive pedofeatures

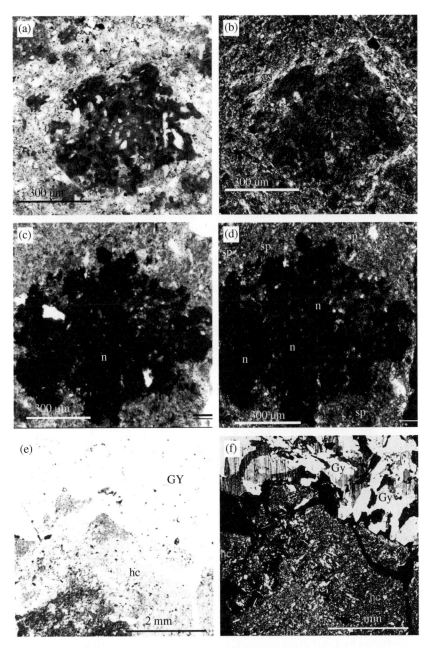

Plate 8.1. *Impregnative and depletion pedofeatures*. (**a**) weakly impregnated orthic Fe- hydroxide nodule in groundmass of yellowish clay and fine sand size angular quartz grains with double spaced to open porphyric c/f related distribution (PPL); (**b**) same, notice granostriated (gs) b-fabric (XPL); (**c**) moderately impregnated orthic aggregate Fe- hydroxide nodule (n) (PPL); (**d**) same, stipple speckled b-fabric (sp) in groundmass surrounding the nodule (n) (XPL); (**e**) calcite depletion hypo-coating (hc) (yellow) around coarse gypsum nodule (GY) (top) and remaining calcitic groundmass (bottom center) (PPL); (**f**) same, notice calcitic crystallitic b-fabric in groundmass and a stipple speckled one in depleted zone (XPL).

to develop. Impregnative nodules that are inherited from the parent material (e.g., transported nodules) should of course not be described as pedofeatures (see Section 8.6.2), but as coarse constituents of the groundmass. It is important to report the degree of impregnation of these pedofeatures. The following scale of purity is proposed (Fig. 8.1):

- *weakly impregnated:* characteristics of the original groundmass are dominant (e.g., size, sorting, and distribution of the coarse material) (Plate 8.1a and b);

- *moderately impregnated:* characteristics of the original groundmass are common and clearly identifiable (Plate 8.1c and d);

- *strongly impregnated:* the material of the pedofeature is dominant, but the nature of the groundmass can still be identified (Plate 8.14f).

If the pedofeature is not related to the groundmass, it is considered an intrusive pedofeature, such as clay coatings. Estimation of the absolute degree of impregnation is impossible, as it is strongly influenced by the nature of the impregnating compound. For example, very small amounts of Mn-(hydr)oxides will give the impression of a strong impregnation.

Depletion pedofeatures: recognizable because of a lower concentration of a component, mainly in the micromass, such as iron, calcite, or clay. Again the term depletion as used here does not refer to a process, but only to the observation that a zone contains less of a given substance than the surrounding groundmass as in the case of a local reduction of Fe oxides, resulting in a gray zone in a yellowish or red groundmass (Plate 8.1e and f, Plate 8.10a and b, Plate 8.21c through f).

Fabric pedofeatures are recognizable from the groundmass because of a difference in fabric only. The most common examples are passage features infillings with groundmass material resulting from bioturbation by soil mesofauna (Plate 8.2a and b, Plate 8.10c and d, Plate 8.21c through f), (See Section 8.4.3). Other examples are compaction pedofeatures, for instance at the soil surface or as hypocoatings around channels (Plate 5.5b. Although some striated b-fabrics could be considered as fabric pedofeatures, they are excluded by definition.

The boundaries of matrix pedofeatures are quite often, but not necessarily, gradual or diffuse. As several characteristics may be superposed (e.g., depletion and fabric features), the key proposed in Table 8.1. has to be followed strictly.

> *Intrusive pedofeatures* are pedofeatures that do not enclose groundmass material, such as large pedogenic crystals, clay coatings, and relative pure nodules of Fe hydroxides. (Stoops, 1998).

Intrusive pedofeatures are either formed in preexisting voids, or in the groundmass by total displacement or replacement of the other constituents, as is frequently the case with calcitic nodules. Sometimes it can be difficult to distinguish them from inherited coarse constituents, such as euhedral crystals derived from the parent material. Intrusive pedofeatures always have sharp boundaries.

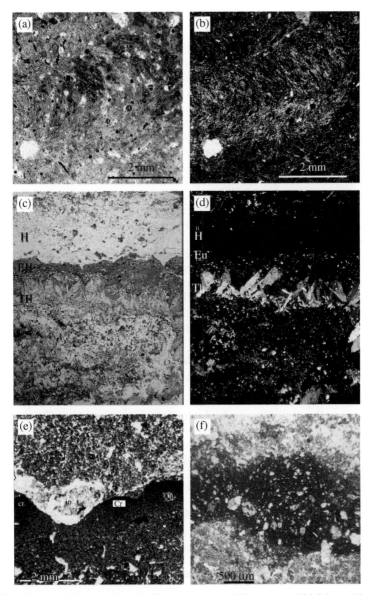

Plate 8.2. *Fabric pedofeatures and crusts.* (**a**) passage feature visible by crescent fabric in grayish ground-mass with double spaced and open porphyric c/f related distribution pattern; small circular air bubbles in mounting medium (a) (PPL); (**b**) same, the passage feature becomes much clearer due to clay orientation (XPL); (**c**) compound surface crust; from top to bottom: halite (H), eugsterite (Eu), thenardite (Th) (PPL); (**d**) same, notice isotropic nature of halite (XPL); (**e**) internal crust (cr) on palaeosurface in loess profile; notice that the grain size becomes finer towards the top of the crust; several infillings and passage features point to a high biological activity (PPL); (**f**) thin Fe-humus pan (PPL).

Volume percentages of intrusive pedofeatures can be estimated if the features are relatively pure (e.g., gypsum crystals, fine clay coatings, goethite nodules), in contrast to the quantification of matrix pedofeatures in volume percentage, which are ambiguous with respect to their interpretation (Stoops, 1977, 1978b, see also Section 4.3.5).

Brewer (1964a) subdivided pedological features into orthic (formed in situ) and inherited. The orthic features correspond to the pedofeatures as defined by Bullock et al. (1985). They were subdivided into plasma concentrations, resulting from a significant change in composition or concentration of any of the fractions of the plasma, and plasma separations, resulting from a significant change in the arrangement of the constituents. The latter thus corresponds to fabric pedofeatures in our text. Brewer subdivided inherited pedological features into lithorelicts (rock fragments), pedorelicts (features formed by erosion, transport, and deposition of nodules of an older soil material, or by preservation of some part of a previously existing soil horizon within a newly formed horizon) and sedimentary relicts (features formed during deposition of a transported soil parent material, such as stratifications). Inherited pedological features have no equivalent in Bullock et al. (1985) because rock fragments, lateritic nodules, etc., are considered as part of the coarse material, and therefore as part of the groundmass and sedimentary features described as orientation and distribution patterns in the groundmass.

8.2.3 Morphological Classification of Pedofeatures as Related to their Fabric

Five main groups of pedofeatures can be distinguished according to their morphology, crystallinity, and related distribution pattern. On the highest level, a distinction is made between pedofeatures that are related to voids (other than packing voids) and natural surfaces of grains, aggregates, or voids, and those that are unrelated to voids or natural surfaces.

1. Related to Natural Surfaces of Voids, Grains and Aggregates

1.1. Coatings, hypo-coatings and quasi-coatings: related to natural surfaces of grains, aggregates or voids

1.2. Infillings: pedofeatures, other than coatings, related to voids

2. Unrelated to Voids, Grains or Aggregates

2.1. Crystals and crystal intergrowths: crystals, mostly euhedral or subhedral, assumed to be formed in situ in the groundmass

2.2. Nodules: equant to prolate pedofeatures occurring in the groundmass

2.3. Intercalations: elongate, undulating pedofeatures, not consisting of crystals or crystal intergrowths

Two or more pedofeatures may occur in combination, forming a *compound* or a *complex* pedofeature (see Section 8.9). Due to a disturbance of the soil, pedofeatures may be *fragmented* and/or *deformed* (see Section 8.10).

8.3 COATINGS, HYPOCOATINGS AND QUASICOATINGS

8.3.1 Introduction

Coatings, especially clay coatings, were among the first pedogenic features to be recognized in thin sections. Their recognition has been a very important criterion in several soil classification systems, especially in U.S. Soil Taxonomy (Soil Survey Staff, 1975), where they are used to define argillic horizons. In the early Russian research, attention was also already given to clay films (e.g., Minashina, 1958). Coatings, hypocoatings, or quasicoatings related to macrovoids are important diagnostic features in soil genesis because they are always formed in situ; however, they can be inherited from an earlier phase of pedogenesis.

8.3.2 Definitions

The following definitions, based on Bullock et al. (1985) and adapted by Stoops (1998) are used:

> *Coatings* are intrusive pedofeatures coating a natural surface (of voids, grains or aggregates) in the soil. Coatings of voids should occupy less than 90% of the original void space.
> *Hypocoatings* are matrix pedofeatures referred to a natural surface in the soil and immediately adjoining it.
> *Quasi-coatings* are matrix pedofeatures referred to natural surfaces in the soil, but not immediately adjoining the natural surfaces.

The difference between coatings, hypo-coatings and quasi-coatings is illustrated in Fig. 8.2. Being intrusive pedofeatures, coatings have sharp boundaries (e.g., clay coatings) and generally show a distinct contrast with the groundmass. Embedded grain coatings are surrounded by the groundmass (Plate 8.8c), whereas free grain coatings cover grains surrounded by voids. Hypocoatings may be

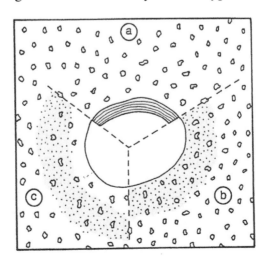

Fig. 8.2. Difference between coating (a), hypocoating (b), and quasicoating (c) around a void such as a channel. Composition of the pedofeatures varies.

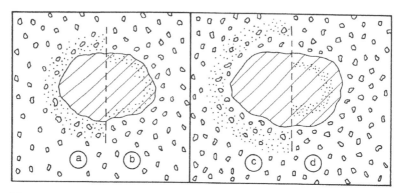

Fig. 8.3 External (a) and internal (b) hypo-coating and external (c) and internal (d) quasi-coating.

Table 8.2. Key to the morphological types of pedofeatures (Stoops, 1998, modified).

1. Matrix pedofeature	
1.1 Associated with the surface of peds, voids or other fabric units and occurring immediately at that surface:	hypo-coatings
1.2 Other matrix pedofeatures related to surfaces but occurring at some distance from the surface:	quasicoatings
1.3 Other strongly elongated or undulating matrix pedofeatures unrelated to surfaces:	intercalations
1.4 Other matrix pedofeatures unrelated to surfaces:	matrix nodules
2. Intrusive pedofeatures	
2.1 Associated to surface of peds(†), voids (but filling them for < 90 %) or the surface of coarse fabric units:	coatings
2.2 Other intrusive features, totally or partially filling voids, other than packing voids:	infillings
2.3 Other intrusive pedofeatures unrelated to surfaces and voids, consisting of euhedral or subhedral single crystals or of intergrowths of such crystals:	crystals and crystal intergrowths
2.4 Other strongly elongated and undulating intrusive pedofeatures unrelated to surfaces and voids:	intercalations
2.5 Other intrusive pedofeatures unrelated to surfaces and voids:	nodules
(†) Ped surfaces have priority over void surfaces.	

of an impregnative, depletion, or fabric type. When developed in the matrix around a grain or aggregate, they are said to be external; when developed in the surface zone of a microporous grain or aggregate they are considered internal (Fig. 8.3, Plate 3.5a through c, Plate 6.4e and f).

Quasicoatings are comparable to hypocoatings with respect to all characteristics, except for the fact that they are not immediately adjoining the surfaces to which they are referred, but occur at some distance. The so-called "Liesegang rings" are an example of quasi-coatings (Plate 8.8d, Plate 8.12a and b).

Coatings may grade into infillings, either because they fill the void more and more, or because of a more oblique or tangential section (Fig. 3.3c and Fig. 8.9., Plate 8.6a-c, Plate 8.8e and f, Plate 8.10e and f);

see also Section 8.4). Tangential sections through hypocoatings may resemble matrix nodules (see also Section 3.1.2).

8.3.3 Classification

Coatings, hypocoatings, and quasicoatings can be subdivided into seven morphological types (Fig. 8.4):

Crust: thick (more than a few mm thick) (hypo-/quasi-)coating on the soil surface. The term applies to discontinuous features. When a crust is continuous, it is described as a layer or horizon (Plate 8.2b and c). When buried, the crust is found in the profile Plate 8.2e). Types and genesis of surface crusts are discussed by Williams et al. (2018).

Micropan: thick (> 0.5 mm) (hypo-/quasi-)coating occurring (sub) horizontally in the soil, varying significantly in thickness over its length (Plate 8.2f).

Capping: (hypo-/quasi-)coating on top of a free or embedded grain or aggregate (e.g., in permafrost soils) (Plate 8.3a, Plate 8.22a and b).

Link capping: a (hypo-/quasi-)coating lying on top, and supported by two or more free or embedded grains or aggregates (Plate 8.3b).

Pendent: (hypo-/quasi-)coating on the lower surface of a free or embedded grain or aggregate (e.g., gypsum or calcite beard below a pebble) (Plate 8.3c and d).

Crescent coating: coating with elongate crescent shape and internal fabric (especially in the case of a clay coating at the bottom of a void) (Plate 8.4a through d, Plate 8..5a and b). They can also be the result of an oblique section through a typic coating. Often they gradually become infillings (see Section 8.4).

Typic (hypo-/quasi-)coatings: not having one of the characteristics mentioned above; they are approximately regular in thickness. Examples include clay coatings on ped surfaces, calcite hypocoatings on channels, Fe-(hydr)oxide coatings on planes in saprolite (Plate 8.3e and f, Plate

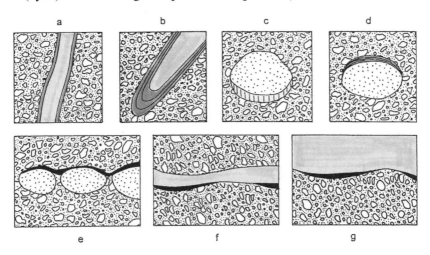

Fig. 8.4. Morphological classification of coatings (a) typic, (b) crescent, (c) pendent, (d) capping, (e) link capping, (f) micropan, and (g) crust (after Bullock et al. 1985).

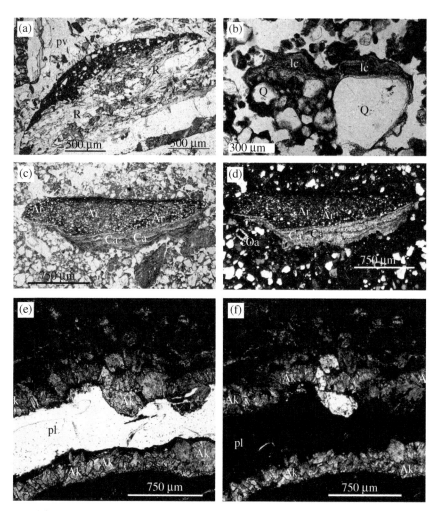

Plate 8.3. *Cappings, pendents and coatings.* (**a**) capping of fine sand, silt and clay on metamorphic rock fragment (R) and packing void (pv) (PPL); (**b**) link capping (lc) of calcite covering several grains; quartz grains (Q) (PPL); (**c**) fragment of fine grained sandstone (Ar) with multi-layered calcite pendent (Ca) (PPL); (**d**) same, notice coating of cytomorphic calcite in channel (cOa) (XPL); (**e**) typic coating of ankerite (Ak) on plane (pl) (PPL); (**f**) same (XPL).

8.8e and f). As the majority of (hypo/quasi-)coatings are typic, the word typic is usually not mentioned in descriptions.

BACKGROUND – The micromorphological concept of coatings was introduced by Brewer (1960) who created the terms cutans, neocutans and quasicutans, which correspond with coatings, hypocoatings, and quasicoatings in Bullock et al. (1985). Adding the suffix-an to a material name, Brewer (1964a) coined a series of terms to designate different types of cutans. Examples include argillan (clay coating), ferri-argillan (reddish or yellowish clay coating), (neo)mangan (Mn (hydr)oxide (hypo)coating), (neo)calcitan (calcite [hypo]coating), neoskeletan (more sandy hypocoating resulting from a loss of fine fraction), neostrian (orientation of clay along a surface, such as in grano- or porostri-ated b-fabric). Neo- and quasi-cutans as used by Brewer (1964a) refer only to the external types. The term "cutan" is also used in field descriptions, especially in the USDA system, in a very restricted sense, for clay coatings.

Cohesive coatings of pure coarse crystalline material are described in the system of Brewer as crystal tubes (in channels), crystal chambers (in vughs, vesicles, and chambers) and crystal sheets (in planar voids). They are subdivisions of the main group of crystallaria. The terms clay coatings, clay films, and clay skins were used earlier in micromorphological descriptions. Kubiëna did not make a distinction between clay coatings and striated b-fabrics.

8.3.4 Descriptive Criteria

The list of coatings discussed below is not exhaustive. For cases not included here, refer to the general instructions on description given in Section 4.3.

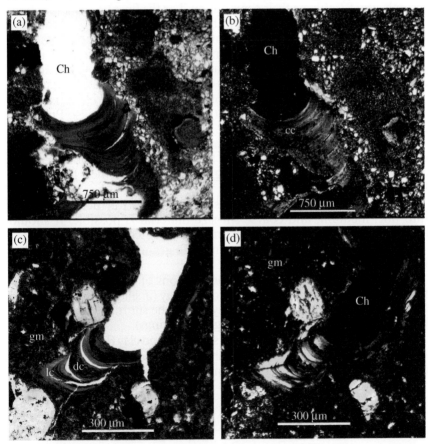

Plate 8.4. *Crescent clay coatings.* (**a**) crescent coating of coarse clay in channel (Ch) (PPL); (**b**) same, the high interference colors point to the presence of 2/1 clays in the coating, and the absence of sharp extinction lines is characteristic for coarse clay coatings (cc) (XPL); (**c**) layered crescent coating of fine limpid reddish clay (lc) alternating with coarse brown clay in channel (dc) in dark groundmass (PPL); (**d**) same, extinction bands, are visible only in the fine strongly oriented limpid clay; groundmass (gm) with undifferentiated b-fabric (XPL).

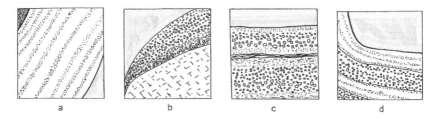

Fig. 8.5. Internal fabric of clay and silt coatings. (a) microlamination, (b) lamination, (c) layering, and (d) compound layering (after of Bullock et al. 1985).

Textural coatings are those characterized by a difference in grain size compared with the groundmass. They are often, but not always, the result of illuviation. Coatings resulting from the transport in suspension of fine material are an important record of current and/or past pedogenic processes. Most of them are composed of size fractions beyond the resolution of the optical microscope, and a detailed observation of their limpidity and grain size is the only way to evaluate their composition. For the finer fraction, the abundance of contrasted microparticles, with a size of 2 to 5 μm is a good criterion. Bullock et al. (1985) proposed the following classes to describe such features:

Limpid clay coatings: uniform fine clay without inclusions of microparticles. If composed of phyllosilicates they generally show interference colors; they indicate stable conditions on and in the soil (Plate 8.5a and b).

Dusty clay coatings: composed of fine to coarse clay comprising microparticles of up to 3 μm diameter (Plate 8.4a and b; see also Plate 8.5c and d).

Impure clay coatings: clay containing numerous contrasted particles of fine silt size. In XPL, mica flakes, if present, can be seen, but quartz particles are less easily recognized because of their lower interference colors (Plate 8.5e and f).

Silt coatings: consisting of silt-sized particles. Further subdivision into fine, medium, and coarse silt-size material is possible. It should also be noted whether or not the grains are coated with clay.

Sand coatings: consisting of sand-size grains. Further subdivision is possible as for silt above.

Clay and silt coatings: variable proportions of silt- and clay-size particles. The class can be subdivided into silty clay (if silt grains are dispersed throughout a clayey mass) and clayey silt (if the silt fraction is dominant but more clay is present than just as grain coatings).

Unsorted: variable proportions of sand, silt, and clay (Plate 8.20c and d).

Clay and silt coatings are probably the most commonly described features, but nevertheless only little systematic information is available on their

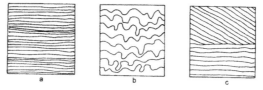

Fig. 8.6. Basic orientation of laminae in clay coatings: (a) parallel, (b) convolute, and (c) cross-laminated (from Bullock et al., 1985).

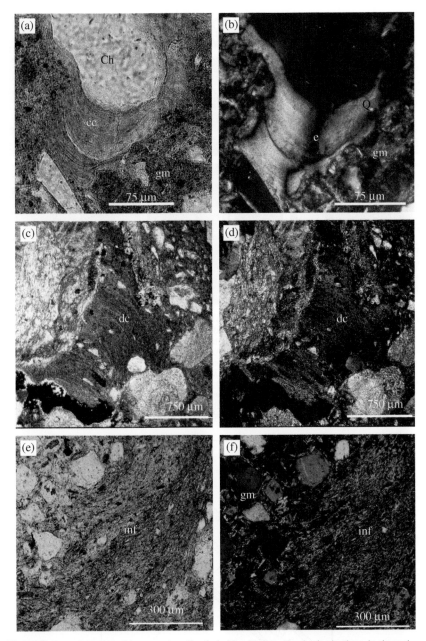

Plate 8.5. *Textural coatings.* (a) crescent coating (cc) of limpid clay with microlaminations, in channel (ch); notice faint contrast with speckled groundmass (gm) (PPL) (b) same, notice increased contrast compared to PPL and clear extinction lines (e) (XPL); (c) dusty clay infilling (dc) (PPL); (d) same, notice absence of clear extinction lines (XPL); (e) silty infilling (inf) of clay with sericite flakes (PPL); (f) same, notice absence of extinction lines (XPL).

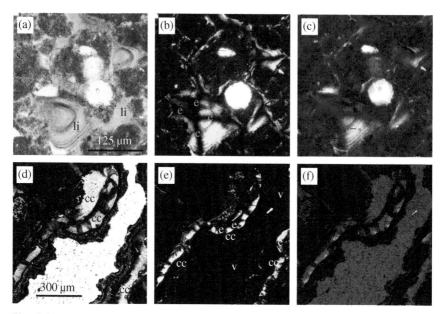

Plate 8.6. *Orientation of clay particles in coatings and infillings.* (**a**) Infillings of yellowish fine clay with continuous orientation; some parts are limpid (li), other speckled (PPL); (**b**) same, notice sharp extinction lines (e), forming a cross in the infilling; the low interference colors points to a kaolinitic composition (XPL); (**c**) same, but with retardation plate ($n\gamma$ oriented NW-SE): addition of interference colors when $n\gamma$ of the retardation plate is parallel to the orientation of the clay particles (blue interference colors (+)), and subtraction when $n\gamma$ is perpendicular to it (yellow interference colors (-)), (XPLλ); (**d**) compound juxtaposed coatings (cc) of limpid yellowish clay in weathered andesite boulder (PPL); (**e**) same, the sharp extinction lines (e) indicate a strong orientation of the fine clay particles. Notice that the layer (cc) close to the void (v) is practically isotropic (XPL); (**f**) same, but with retardation plate ($n\gamma$ oriented NW-SE): the yellow interference color of the coating is a result of subtraction, and proves that the clay particles are oriented perpendicular to the walls (XPLλ).

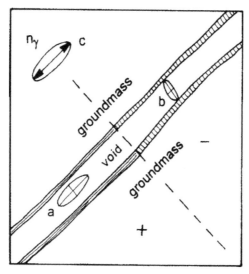

Fig. 8.7. Use of ℓ retardation plate (gypsum compensator) to demonstrate the difference between illuvial (left) and newly formed (right) limpid clay coatings. In the first case, when the clay is oriented parallel to the wall, interference colors will become of a higher order, in the latter case, when the clay is oriented perpendicular to the wall, interference colors of a lower order will appear.

internal fabric, even though it can contain crucial information (e.g., Fedoroff, 1997; Kühn et al., 2018). The following types of internal fabrics are distinguished (Fig. 8.5):

Nonlaminated: the coating appears as a homogeneous body. This term can be applied to coatings of all grain size classes.

Microlaminated: alternating thin (< 30 µm) laminae of limpid and speckled clay (Plate 8.5a and b, Plate 8.6a through c, Plate 8.22c and d).

Laminated: alternating thick (> 30 µm, usually 100–200 µm) laminae with the same texture but with differences in color or limpidity.

Layered: alternating layers with different textures (Plate 8.4c and d).

Compound layered: alternating layers of microlaminated clay and clay with silt.

The thickness and the related distribution of the laminae (parallel, convolute, cross-laminated) should also be indicated (Fig. 8.6).

The presence of extinction lines is an important criterion for determining the degree of internal orientation in clay coatings and their interpretation (see also Section 4.3.2.2.3).

Sharp extinction lines point to a strong continuous (parallel) orientation of the clay particles (Plate 3.2a, Plate 8.5b; Plate 8.6b). This is observed generally only in limpid fine clay (< 0.2 µm) coatings.

Diffuse extinction lines: clay particles are only partly oriented.

Absence of extinction lines: most of the clay particles are not parallel oriented, or amorphous material is dominant.

For sand, silt, and unsorted coatings, it is important to describe the internal fabric based on grain-size distributions (e.g., fining up and coarsening). Some Fe-stained fine clay coatings show clear pleochroism. Clay coatings may show alteration related to hydromorphic conditions, whereby they gradually lose their anisotropy, become grainy and white opalescent in OIL (Brinkman et al., 1973).

Illuvial and newly formed nonlaminated clay coatings are generally distinguishable based on their optical orientation: In illuvial coatings, the clay particles are oriented parallel to the walls of the void, and behave as length slow bodies (Plate 4.3, Plate 8.6a through c), whereas with in situ neoformed clay coatings, the clay particles are generally oriented perpendicular to the walls and behave as a whole as length fast elongated bodies, showing lower interference colors when the retardation plate is inserted (Fig. 8.7; Plate 8.6d through f; see Section 4.3.2.2.3). Radial sections through fresh straw can sometimes be confused with nonlaminated limpid yellow clay coatings on channels (Plate 8.7e through g).

Limpid isotropic coatings are in most cases composed of allophane or allophane-like materials, but similar coatings consisting of ferrihydrite and phosphates, such as calcioferrite (Jenkins, 1994) have also been described (Plate 8.7a through d).

Clay and silt or unsorted accumulations on top of gravels (Plate 8.3a) or peds (Plate 8.22a and b) that form cappings and/or link cappings are a widespread feature in soil horizons subject to repeated freezing and thawing. They have been described as silt droplets, silt cappings and banded fabrics (FitzPatric, 1956; Dumanski and St. Arnaud, 1966; Romans et al., 1966; Van Vliet-Lanoë and Fox, 2018) (see also Section 8.9).

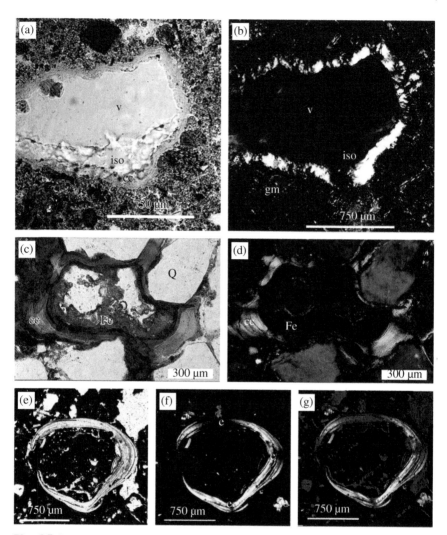

Plate 8.7. *Compound coatings of clay and amorphous material.* (**a**) compound typic coating of yellow-ish limpid, strongly oriented clay (cc) on vugh (v), in the bottom part covered by a juxtaposed coating of limpid colorless isotropic material (iso) (PPL); (**b**) same, notice the strong continuous interference colors of the clay coating with extinction bands, and the absence of interference colors in the isotropic coating (iso) (XPL); (**c**) compound juxtaposed coating of clay (cc) and of amorphous Fe-hydroxide (Fe) (PPL); (**d**) same, notice isotropic appearance of the Fe- hydroxide (Fe) and the extinction lines in the clay coating (cc) (XPL); (**e**) section through a straw filled by groundmass material; compare this with a coating of fine clay (PPL); (**f**) same, notice undifferentiated b-fabric in the groundmass and extinction bands (e) in the straw (XPL); (**g**) same, notice addition and subtraction of interference colors (XPLλ).

Impure clay coatings and infillings, containing high amounts of small organic particles (punctuations) and stained by organic material are found for instance in agric horizons and in archaeological contexts, but also in natural soils (Deák et al., 2017).

Coatings consisting of sand grains should not be confused with clay depletion hypocoatings, as described below. Two criteria can be used to distinguish both types:

1. Boundaries of coatings are sharp, whereas those of depletion hypocoatings are generally diffuse.

2. Composition and texture of the sand is identical with that of the groundmass in the case of depletion hypocoatings, but can be different in the case of coatings.

Illuvial clay coatings and pressure faces are often difficult to distinguish in the field. Micromorphology is the best method to solve this problem.

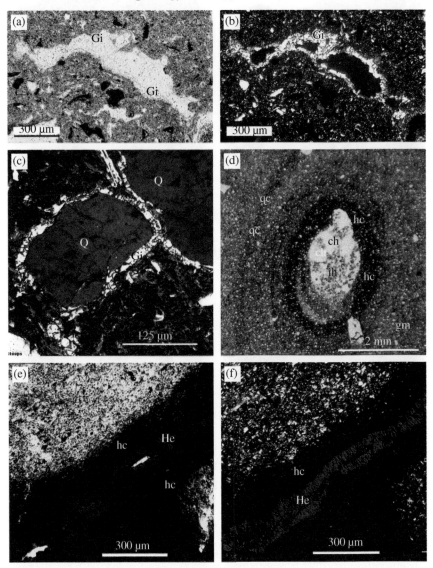

Plate 8.8. *Crystaline coatings*. (**a**) typic coating of fine crystalline gibbsite (Gi) on prolate vugh (PPL); (**b**) same, the gibbsite coating shows a xenotopic fabric when observed at higher magnification (XPL); (**c**) typic embedded grain coatings of gibbsite on rounded quartz grains (Q) (XPL); (**d**) hypo- (hc) and quasi-coating (qc) of Fe- hydroxide (so called Liesegang rings) around channel (ch) with loose discontinuous infilling in sandstone (PPL); (**e**) coating and infilling of hematite (He) oriented perpendicularly to the wall of the plane, and hypo-coating (hc) of Fe- and Mn- hydroxides in saprolite (PPL); (**f**) same (XPL).

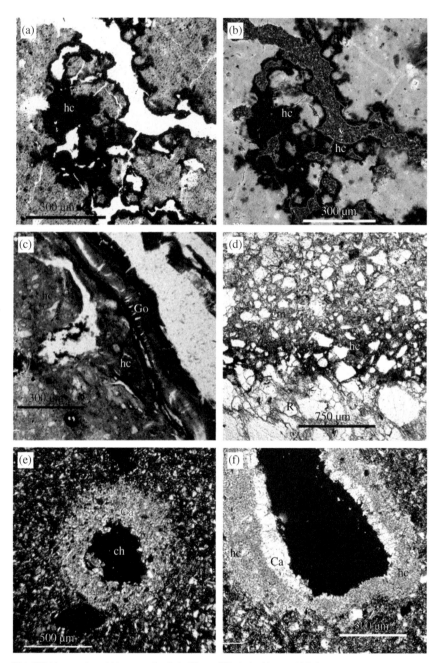

Plate 8.9. *Hypocoatings.* (**a**) hypo-coating (hc) of Fe- and Mn- hydroxide on void in homogeneous, speckled groundmass with open porphyric to fine monic c/f related distribution pattern (PPL); (**b**) same, the presence of iron is shown by the brown color (TDFI); (**c**) coating of fibrous goethite (Go) oriented perpendicularly to the wall of the channel, and hypo-coating (hc) of fine goethite (PPL); (**d**) coarse rock fragment (R) (lower half of the micrograph) with hypo-coating (capping) of amorphous organic fine material in groundmass with close porphyric c/f related distribution pattern (PPL); (**e**) calcite hypo-coating on channel (ch) in loess (pseudomycelium according to field description) (XPL); (**f**) channel covered by typic, coarse calcite coating (Ca) and surrounded by micritic hypo-coating (hc); the groundmass has a calcitic crystallitic b-fabric (juxtaposed compound pedofeatures) (XPL).

In PPL, intrusive clay coatings can be identified by their contrast with the groundmass, their sharp boundaries, and their microlaminations. Pressure faces are not visible in PPL. In XPL, illuviation coatings show a continuous orientation of the clay particles parallel to the aggregate walls, generally emphasized by clear extinction lines; pressure faces appear as porostriated b-fabrics, with diffuse boundaries. In XPL, clay coatings deformed by stress may be identical to pressure faces, but remain visible in PPL by their contrast. For more information on textural features see also Kühn et al. (2018).

Crystalline coatings are found under different pedogenic conditions, and include for instance calcite coatings (Plate 8.3e and f) in semiarid soils, goethite coatings in hydromorphic soils (Plate 8.7c and d), and gibbsite coatings in tropical soils (Plate 8.8a through c;), and hematite coatings in lateritic saprolites (Plate 8.8e and f). Their description comprises crystal size, shape, orientation, and distribution (e.g., parallel distribution of layers of acicular crystals oriented perpendicular to the walls) (Plate 4.1e and f). In the case of relatively thick coatings, the terminology given for crystalline nodules (Section 8.6.4.) can be used to describe the internal fabric. A so-called *palisade fabric*, composed of parallel prismatic crystals oriented perpendicular to the covered surface, is often observed.

BACKGROUND – Crystalline coatings partly cover Brewer's concept of crystal tubes and crystal sheets as subdivisions of his crystallaria.

Calcite (Ducloux et al., 1984) and gypsum often form pendents below pebbles (Plate 8.3c and d) the latter especially in gravelly terrace materials. The absolute spatial position of the pendent can yield information on a possible disturbance of the soil. Gypsum and more soluble salts form surface crusts in arid areas (Plate 8.2c and d) (e.g., Driessen and Schoorl, 1973, Vergouwen, 1981; Mees and Stoops, 1991; Mees and Tursina, 2018).

Coatings of cryptocrystalline and amorphous materials require a description of their nature and any observed internal fabric (Plate 8.7a through d)). Identification of the composition will in most cases only be possible after additional studies (e.g., UVF, BLF, selective chemical extraction, staining, WDS). In the description of coatings of organic matter, it is important to distinguish monomorphic and polymorphic materials (see Section 6.4.3).

Impregnative hypocoatings and **quasicoatings** mainly consist of amorphous (e.g., organic matter), cryptocrystalline (e.g., Fe and/or Mn (hydr) oxides) or microcrystalline (e.g., calcite) material (Plate 8.8d through f; Plate 8.9a through c) occurring in a higher concentration than in the groundmass. Description should include an estimation of the degree of impregnation (see Section 8.2; Fig. 8.1). Impregnative hypocoatings can also be formed artificially by percolating the soil with a dye such as methylene blue (to detect the conducting pores) (Plate 3.6d). Microcrystalline hypocoatings generally occur on voids, mainly channels (Plate 8.9e and f). Coarse crystalline hypocoatings (e.g., calcite) exist, but are rather rare and generally show a poikilotopic fabric (see Section 8.6.4). In soils with a calcareous micromass, the borders of aggregates and voids may show more intense interference colors due to wedging effects (see Section 3.1.3); these linings should not be mistaken for hypocoatings.

Mn-(hydr)oxide hypocoatings, recognizable by their opacity, point to weak hydromorphic conditions (Plate 8.9a and b); Fe-(hydr)oxide hypocoatings/quasicoatings are significant indicators for redoximorphic conditions (Veneman et al., 1976; Vepraskas et al., 1994; Vepraskas et al., 2018) (Plate 8.7c and d; Plate 8.9c). Porous gravels (e.g., some fine sandstones and cherts) can under similar conditions exhibit internal Fe-hypocoatings. Capping hypocoatings of monomorphic organic material have been described in tropical podzols (Plate 8.9d).

Depletion hypocoatings are generally the result of removal of part of the micromass, either in solution (e.g., calcite) or in suspension (e.g., clay). Iron depletion hypocoatings are sometimes observed in hydromorphic soils as a result of reduction; they were called albans, following Brewer's terminology, by Veneman et al. (1976) (Plate 8.10a and b). Calcite depleted hypocoatings are frequently observed around roots in calcareous Mediterranean soils (see also Section 8.10, complex pedofeatures) (Plate 8.21c through e) and near petrogypsic zones in aridisols (Plate 8.1e and f). Leaching of the finer fraction of the groundmass leads to coarse-grained depletion hypocoatings around conducting voids, as observed in some E-horizons. They correspond to the neoskeletans of Brewer (1964a). They also occur in washout layers of surface crusts (West et al., 1992; Chiang et al., 1994; Williams et al., 2018). It should be remembered that depletion, as used here, does not point to a process, but to a relative concentration. A depletion can therefore be the result of an accumulation in the surrounding area.

Fabric hypocoatings generally result from mechanical forces, either reorienting the components of the groundmass, or compacting them (Plate 5.5b). If the orientation affects only the fine fraction, the result will not be visible in PPL, but only in XPL, and the feature will key out as a striated b-fabric, rather than a pedofeature. Compaction may lead to a more massive zone, such as around a channel where the enaulic or gefuric c/f related distribution of the groundmass is altered to a close porphyric c/f related distribution pattern. Some crusts and pans also belong to this type.

8.4 INFILLINGS

Well-known examples of infillings on macroscale are the so-called crotovina (also krotovina) and frost wedges. In many cases, infillings form as a result of biological activity.

8.4.1 Definition

Infillings are (previous) voids, other than packing voids, filled or partly filled by soil material or some fraction of it (after Bullock et al. (1985), modified).

The definition by Bullock et al. (1985) does not explicitly take into account infillings by gefuric or chitonic–gefuric material. This has been overcome in the classification given here. The definition cited above is more restrictive than that of the Bullock et al. (1985), as infillings of packing voids are excluded from the concept, because they are considered to be part of either the groundmass or as impregnative pedofeatures.

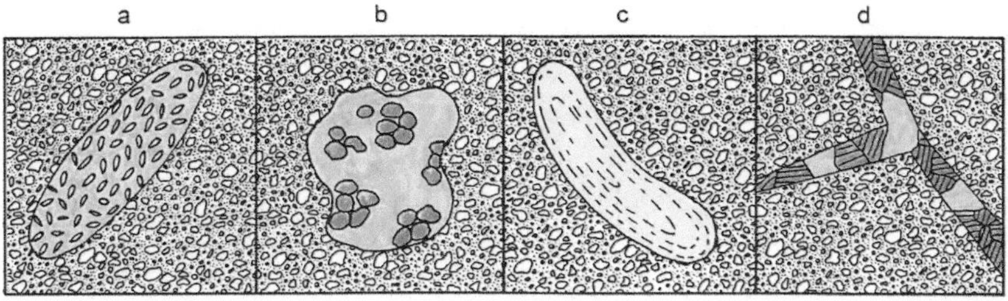

a   b   c   d

Fig. 8.8. Types of infillings: (a) loose continuous (e.g,. lenticular gypsum in channel), (b) loose discontinuous (e.g. excrements in vugh), (c) dense complete (e.g. illuvial clay in channel), and (d) dense incomplete (e.g. calcite in plane).

To be classified as a dense infilling rather than a coating, the void must be more than 90% filled.

BACKGROUND – The concept of infillings, as described Bullock et al. (1985), is broader than that of pedotubules (Brewer & Sleeman, 1963; Brewer, 1964a), the latter being restricted to channel and chamber infillings, and excluding crystalline and clayey features. Only part of channel infillings fit in their concept of pedotubules, which are pedological features consisting of soil material (skeleton grains or skeleton grains plus plasma) with a tubular external form. Accumulations of pure plasma (e.g., clay, calcite, and gypsum) were excluded. Four subtypes were distinguished: granotubules, filled with skeleton grains; aggrotubules, filled with randomly organized aggregates of skeleton grains and plasma; isotubules, filled with a material with a porphyric related distribution and random arrangement, and striotubules, also with a porphyric-related distribution but with a crescent-like internal fabric. Three genetic groups are distinguished: orthotubules, having the same composition as the surrounding soil material, metatubules filled with material derived from another horizon of the same profile, and paratubules with material unlike that of any horizon in the profile. Both series of terms could be combined, for example, ortho-aggrotubules or meta-striotubules.

8.4.2 Classification

Infillings can be classified according to the type of void that has been filled, and the nature of the infilling material. Four main types are distinguished based on fabric and homogeneity (Fig. 8.8):

Dense complete: void is completely filled, and no voids occur within the infilling.

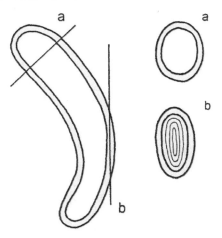

Fig. 8.9. Transversal section through clay coating (a) and tangential section simulating a clay infilling (b).

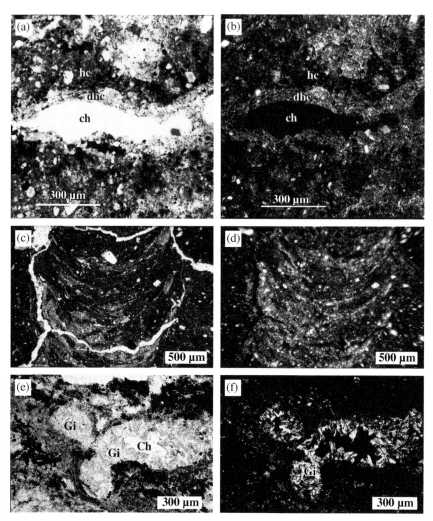

Plate 8.10. *Depletion coating* (**a**) Fe-hypocoating (hc) and superposed Fe-depletion hypo-coating (yellow) (dhc) on channel (ch) (PPL); (**b**) same, notice that the interference colors are stronger in the depleted zone (XPL); *Infillings.* (**c**) dense complete infilling with crescent fabric in channel; compare with crescent illuviation coating where layers are more continuous (PPL); (**d**) same, notice that not only the distribution, but also the orientation of the fine particles is crescent-like (XPL); (**e**) channel (ch) coating and infilling of gibbsite (Gi) (PPL); (**f**) same, notice the euhedral or subhedral shape of the gibbsite crystals (XPL).

Dense incomplete: void is completely filled, but some voids are present within the infilling; if a coating occupies > 90% of the void space, it is considered as an infilling.

Loose continuous: loosely packed material, continuously distributed.

Loose discontinuous: loosely packed material grouped in clusters.

8.4.3 Descriptive Criteria

Dense infillings include for instance the following types:

Infillings of (illuvial) clay; dense complete or incomplete infillings resulting from an overdevelopment of clay coatings, occupying more than 90% of the pore space. Their internal fabric is quite often characterized by a parallel or crescent arrangement of the clay (Plate 8.5c through f, Plate 8.6a through c). For their description, see Section 8.3.4. They are found mainly in channels and planes. A tangential section through a clay coating covering a bend channel may give the impression of a dense clayey infilling (Fig. 3.3c, Fig. 8.9) (Plate 8.6a through c).

Infillings by crystals: infillings ranging from dense complete to loose continuous, consisting of crystals such as gypsum, calcite (Plate 8.11a), gibbsite (Plate 8.10e and f) or goethite. The internal fabric of the dense infillings should be described as explained in Section 8.6.4. Crystal infillings have been observed in all types of voids (Plate 8.20e and f).

Passage features: are dense complete infillings having approximately the same composition as the groundmass, but can be distinguished on the basis of a more or less pronounced crescent internal fabric resulting from a preferential distribution of the different size fractions or components (Plate 4.1d; Plate 8.2a and b; Plate 8.10c and d). They generally lack distinct boundaries and point to the presence of mobile soil fauna. Their external shape usually corresponds to a channel. They are characteristic features in large termite mounds (Stoops, 1964) and vermicular microstructures (see Section 5.4.2), and are comparable to the ichnofabrics described in sediments (Ekdale and Bromley, 1991).

Loose infillings: formed by a variety of materials including excrements and coarse constituents (Plate 6.5e and f) and granules (Plate 8.11b,). In the latter cases, the term *microgranular infillings* can be used. Note that the infilling material generally consists of other pedofeatures (e.g., excrements), which have to be described as such. The size and the shape of the individual components of the infilling have to be described. The resulting porosity pattern generally corresponds to packing voids, although the global feature must be described for instance as a filled channel or a vugh.

8.5 CRYSTALS AND CRYSTAL INTERGROWTHS

8.5.1 Introduction

Euhedral crystals of pedogenic origin are striking and characteristic features of arid soils (gypsum, calcite), and of reduced wet soils (siderite, pyrite, vivianite). Kubiëna (1938) called them intercallary crystals and used them as a criterion to distinguish several fabric types. If well-described, euhedral crystals can yield important information on environmental conditions at the time of their formation. A popular type of crystal intergrowth is the so-called "desert rose" found in Algeria and

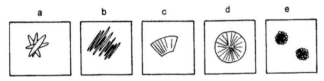

Fig. 8.10. Tentative classification of crystal intergrowths: (a) random, (b) parallel, (c) fanlike, (d) radial, and (e) framboids.

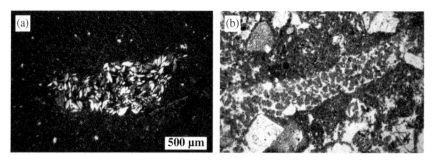

Plate 8.11. *Loose infillings.* (**a**) loose continuous infilling of vugh with lenticular gypsum; stipple speckled b-fabric of groundmass (XPL); (**b**) loose continuous infilling of channel by mineralo-organic excrements (PPL).

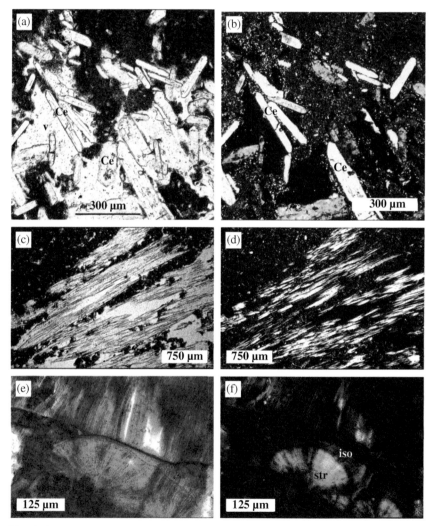

Plate 8.12. *Crystals and crystal intergrowths.* (**a**) prismatic crystals of celestine (Ce) in groundmass and voids (v) (PPL); (**b**) same; notice calcitic crystallitic b-fabric in groundmass (XPL); (**c**) parallel intergrowth of gypsum (fibrous gypsum) (PPL) (**d**) same (XPL); (**e**) fanlike intergrowth of fibrous goethite (PPL); (**f**) same, notice differences in intensity of interference colors grading from strong (str) to isotropic (iso) (XPL)

Tunisia, composed of an intergrowth of lenticular gypsum crystals of several centimeters each.

8.5.2 Definition

> *Crystals and crystal intergrowths* are pedofeatures consisting of single crystals and intergrowths of crystal larger than 20 μm, euhedral or subhedral, embedded in the groundmass, and that are not present as such in the parent material (Bullock et al. (1985), modified).

Pedogenic minerals occurring frequently as crystals or crystal intergrowths include gypsum and calcite, and less commonly celestite, barite, siderite, vivianite, pyrite, and soluble salts.

8.5.3 Classification

In Bullock et al. (1985) no subdivision was proposed for this group. Because of their importance in some soils, a subdivision, based on distribution and/or orientation pattern is proposed here (Fig. 8.10):

Single crystals (Plate 6.2a and b, Plate 8.12a and b).

Crystal intergrowths:

Random: as in gypsum desert roses.

Parallel: as in parallel fibrous gypsum, satin spar (Plate 8.12c and d).

Fanlike: as in fibrous goethite (Plate 8.12e and f).

Radial around a central axis.

Radial around a central point or a short central axis, so called spherulites (e.g., calcite, vivianite) (Plate 8.13a through d).

Framboids: spherical features composed of an agglomeration of smaller spherical particles (e.g., pyrite) (Plate 3.2e and f).

Attention should be taken to avoid confusing crystal intergrowths with dense complete crystal infillings. Spherulites may be also inherited (e.g., biospherulites, Section 6.2.4.4) and are then part of the coarse fraction of the groundmass.

BACKGROUND – In Brewer (1964a), crystals and crystal intergrowths are described as intercallary crystals, a subdivision of crystallaria.

8.5.4 Descriptive Criteria

The most important descriptive criteria are nature and shape. Other criteria include the presence of inclusions (and their orientation and distribution) or dust lines, orientation and distribution patterns and sizes. Terms used for the description of crystalline nodules (Section 8.6.4) can also be used for crystal intergrowths.

It is not always obvious whether euhedral crystals are pedogenic or inherited. If inclusions within crystals are similar to elements of the groundmass, a pedogenic origin is probable. Transported crystals can generally be recognized by their shape (e.g., rounded edges).

8.6 NODULES

8.6.1 Introduction

Nodules in soil materials have been described in the field since the early days of soil science (e.g., in hydromorphic or calcareous soils) and were without doubt the first features to be recognized in thin sections of soil (Kubiëna, 1938).

A specific problem in nodule description is caused by the field concept of hard nodules vs. soft mottles or flecks. Because consistency cannot be observed in thin sections, this criterion is not valid. In micromorphological description, concretions, nodules, mottles, and flecks are therefore treated as one concept, namely as nodules. When the proposed classification system is followed, sufficient information is

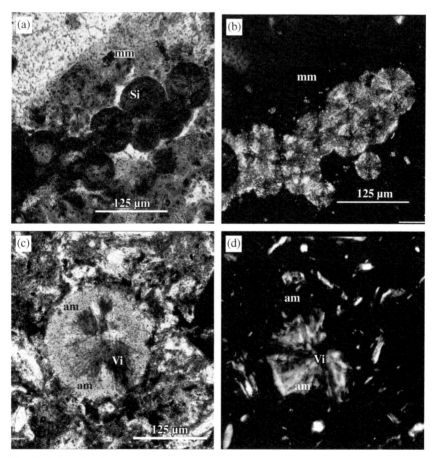

Plate 8.13. *Crystal and crystal intergrowths.* (**a**) spherulites of siderite (Si) in brown organic micromass (mm) (PPL); (**b**) same, notice presence of pseudo-uniaxial figures in the spherulites, and the undifferentiated b-fabric in the micromass (mm) (XPL); (**c**) spherulite of vivianite (Vi) with pellicular alteration to a yellow amorphous material (am) (PPL); (**d**) same, the alteration product is isotropic (XPL).

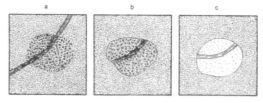

Fig. 8.11. (a) Orthic, (b) disorthic, and (c) anorthic nodules. Orthic nodules formed in place and have not moved. Disorthic nodules formed in the soil where they occur, but they have moved. Anorthic nodules are inherited features that formed in a different location than where they are currently found.

generally recorded to deduce whether a nodule is a hard body or a soft impregnation.

8.6.2 Definition

> Nodules are more or less equidimensional pedofeatures that are not related to natural surfaces or voids and that do not consist of single crystals or crystal intergrowths as defined before (Bullock et al. (1985), modified).

According to their definition, nodules should show no evidence of having formed in a void. Especially for crystalline pure nodules, this may sometimes be hard to prove. For example, a single section through a dense continuous infilling may be mistaken for a nodule, and vice versa. Observing a population of such features generally will give the key for a correct evaluation. Excrements may key out as nodules, but should be distinguished from the latter by their composition and fabric.

It is not always clear whether a nodule is pedogenic and thus a pedofeature, or inherited and therefore a component of the groundmass. Nodules formed in situ are said to be *orthic* (e.g., Plate 8.1a through d; Plate 8.14e and f; Plate 8.15a and b). They can have sharp or diffuse boundaries. When formed in situ, but subjected to local translocations (e.g., rotations in vertic materials) and have sharp boundaries (Fig. 8.11) are called *disorthic*. Nodular bodies with sharp boundaries inherited from the parent material (e.g., in a soil on colluvium or alluvium) are considered as *anorthic* (Plate 8.16b) (Wieder and Yaalon, 1974). However, anorthic nodules may also be of pedogenic origin (e.g., in sediment derived from a soil). Distinguishing orthic and anorthic matrix nodules may prove rather difficult and it is sometimes impossible. A few guidelines are given in Table 8.3.

According to Tardy (1993) it is very important for the interpretation of ferruginous (anorthic) nodules in tropical soils to distinguish between *lithomorphic* nodules, formed in the saprolite and displaying the fabric of the original rock, and *pedomorphic* nodules with a soil fabric.

8.6.3 Classification

8.6.3.1. Internal Fabric

According to their internal fabric, seven different types of nodules were distinguished by Bullock et al. (1985) (i) typic, (ii) concentric, (iii) nucleic, (iv) geodic, (v) septaric, (vi) pseudomorphic and (vii) halo. The classification used in this book is different in several respects: the term halo has

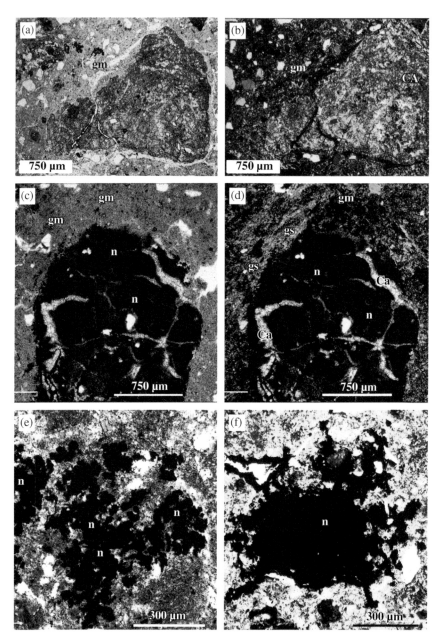

Plate 8.14 *Nodules.* (**a**) typic orthic calcite nodule (fractured) in yellowish gray speckled groundmass (gm) (PPL); (**b**) same, notice presence of quartz grains in the nodule, pointing to an impregnation or substitution (XPL); (**c**) disjointed micritic nodule (n) in yellowish groundmass (gm); the fissures have been filled with calcite (PPL); (**d**) same the presence of coarse crystalline calcite (Ca) in the cracks is clearly seen; notice the granostriated b-fabric (gs) (XPL); (**e**) orthic aggregate Mn- hydroxide nodule (n) (XPL); (**f**) strongly impregnated dendritic Fe- and Mn- hydroxide nodule (n) (PPL).

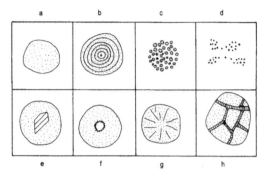

Fig. 8.12. Types of nodules according to internal fabric types (a) typic, (b) concentric, (c) aggregate, (d) dendritic, (e) nucleic, (f) geodic, (g) septaric, and (h) alteromorphic.

Table 8.3. Some criteria to distinguish orthic, disorthic and anorthic matrix nodules.

Characteristic	Orthic	Disorthic	Anorthic
boundary	gradual or sharp	sharp	sharp
internal groundmass	identical to ground-mass of horizon with respect to nature and spatial arrangement †	identical to ground-mass of horizon with respect to nature and spatial arrangement †	generally different from groundmass of horizon with respect to nature and spatial arrangement
internal fabric	continuous	some fabrics may be truncated at the border of the nodule	some fabrics may be truncated at the border of the nodule
location	in groundmass	in groundmass or in an infilling	in groundmass or in an infilling

† An exception has to be made for nodules resulting form so-called epigenetic processes whereby the original minerals are slowly replaced by others, e.g., quartz by calcite (Turc et al. 1985). Microcrystalline nodules are formed containing visibly less quartz grains than the surrounding groundmass.

been deleted, because it belongs to a different level and partly corresponds to hypocoatings. Aggregate nodules have been added as we consider them a type of internal fabric rather than as an external morphology as done by Bullock et al. (1985). The term pseudomorphic has been replaced by alteromorphic based on the considerations presented by Delvigne (1994) and explained in Section 6.2.2. The definition of typic nodules has been widened by deleting the requirement of sharp boundaries.

The following types of nodules are distinguished (Fig. 8.12):

Typic Nodule: nodule with undifferentiated fabric (e.g., rust flecks) (Plate 3.5, Plate 8.1 a through d, Plate 8.14a through d);

Concentric Nodule: nodule with a concentric fabric, such as most manganese concretions in vertic materials (Plate 8.15a through e) and gibbsite nodules in bauxites;

Aggregate Nodule: nodule composed of an aggregation of small, mostly typic nodules. Iron or Mn aggregate nodules are very common in hydromorphic soils (Plate 8.14e). Frequently, the small nodules are distributed according to a dendritic pattern on a mesomorphic scale, hence a subtype of *dendritic nodules* is proposed (Plate 8.14f). Aggregate nodules can have a concentric fabric, but these key out first.

Nucleic nodule: nodule built around a foreign core (Plate 8.16a);

Geodic nodule: nodule with a hollow interior; the void is generally lined by euhedral crystals. This type is very rare in soils. Sections

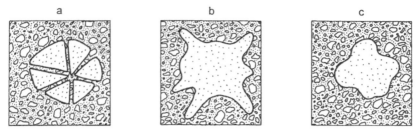

Fig. 8.13. Specific external morphologies of nodules: (a) disjointed, (b) digitate, and (c) mammillate.

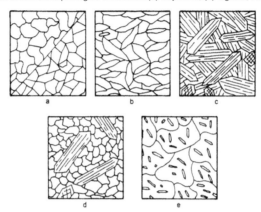

Fig. 8.14. Internal fabric of crystalline pedofeatures. (a) equigranular xenotopic: sutured (top) and mosaic- like (bottom); (b) equigranular hypidiotopic (the shape of the lenticular habit is still visible, although not well expressed); (c) equigranular idiotopic (most grains have their own crystal outlines); (d) porphyrotopic (large euhedral grains enclosed in a mass of finer crystals) and (c) poikilotopic (smaller crystals embedded in larger crystals).

through channels coated with crystalline material may appear as geodic nodules.

Septaric nodule: nodule with a radiating crack pattern; very rare in soils;

Alteromorphic nodule: internal fabric is an alteromorph after some material, for example, a mineral, a rock or a plant fragment (e.g., composed of opal, calcite or goethite) (Plate 8.16b). Alteromorphs after minerals are to be described as part of the coarse material of the groundmass, when they are not formed in situ.

Combinations of characteristics are possible. For example, nucleic nodules are often also concentric.

8.6.3.2. External Morphology

Special external morphologies can be distinguished (Fig. 8.13):

Digitate: fingerlike penetrations of adjacent material.

Disjointed: composed of angular accommodating fragments (Plate 8.14c and d).

Mammillate: outline characterized by several rounded protuberances.

BACKGROUND – Brewer and Sleeman (1964) and Brewer (1964a) defined glaebules as three-dimensional pedological features occurring within the groundmass and not having formed in a void. They partly correspond to the nodules as defined by Bullock et al. (1985). The main differences, pointed out already in Section 7.1, is that rock fragments, inherited nodules and fragments of pedofeatures were considered as glaebules by Brewer (Brewer and Sleeman, 1964), not by Bullock et al. (1985). Six types of glaebules were distinguished by Brewer (1964a): nodules, having an undifferentiated internal fabric, including rock fabrics, concretions with a concentric fabric; septaria with a system of

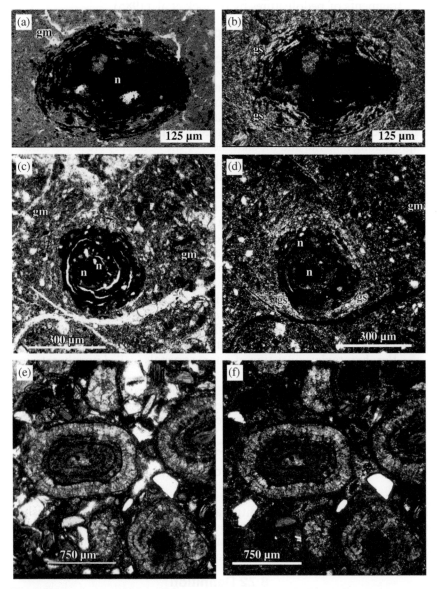

Plate 8.15 *Concentric nodules.* (**a**) concentric impregnative orthic aggregate nodule (n) of Fe- hydroxide in speckled groundmass (gm); notice the fragile outer rings, excluding any transport (PPL): (**b**) same, notice granostriated b-fabric (gs) (XPL); (**c**) orthic concentric impregnative Fe- hydroxide nodule (n) in groundmass (gm) (PPL); (**d**) same, notice granostriated b-fabric (XPL) (**e**) concentric calcite nodules (PPL); (**f**) same, notice radial orientation of calcite crystals (palisade fabric) (XPL).

concentric and radiating cracks; pedodes, with a hollow interior; glaebular haloes, representing weak accumulations surrounding a glaebule with higher concentration of the same constituent, and having diffuse boundaries; and papules, which are clayey glaebules with a continuous or lamellar fabric and sharp boundaries. Papules comprise, for instance, fragments of clay coatings, weathered micas and shale fragments, which are considered as fragments of pedofeatures by Bullock et al. (1985).

8.6.4 Descriptive Criteria

In general a description of the nature, size shape, and boundary is a minimum requirement for the characterization of a nodule. In the case of an anisotropic groundmass, a granostriated b-fabric may point to a relatively greater hardness or consistency of the nodule as compared with the surrounding groundmass (Plate 8.14d; Plate 8.15b and d). In the case of impregnative nodules, a description of the degree of impregnation, the nature of the impregnating material and the sharpness of the boundary are important.

For crystalline nodules and coatings the terminology proposed by Friedman (1965) and adopted by Bullock et al. (1985) for describing the internal fabric is followed here (Fig. 8.14):

Equigranular: the constituent crystals are approximately of the same size; according to their shape three subtypes can be distinguished:

> **Xenotopic**: consisting of anhedral crystals (e.g., commonly observed in coarse calcite nodules (Plate 4.1e and f; Plate 8.16c).

> **Hypidiotopic**: consisting of subhedral crystals (e.g., in hard gypsum infillings) (Plate 8.16d).

> **Idiotopic**: consisting of euhedral crystals (rare).

Inequigranular: the constituent crystals vary in size and frequently show a bimodal size frequency distribution. According to the related distribution of the fine and coarse crystals two subtypes can be distinguished:

> **Porphyrotopic**: consisting of coarse crystals (called porphyrotopes) embedded in a finer mass.

> **Poikilotopic:** larger crystals (called poikilotopes) enclosing smaller elements of a similar (e.g., two generations of gypsum) or different material (e.g., celestite crystals in larger gypsum crystals) (Plate 8.16e and f).

8.7 INTERCALATIONS

8.7.1 Introduction

This concept was first presented by Bullock et al. (1985), but it was not further developed. Little additional information has been published since. The formation of intercalations has not been documented, but several processes might be responsible, such as internal concentration of given size fractions, or neoformation.

8.7.2 Definition

Intercalations are elongate, undulating pedofeatures unrelated to natural surfaces and not consisting of single crystals or crystal intergrowths.

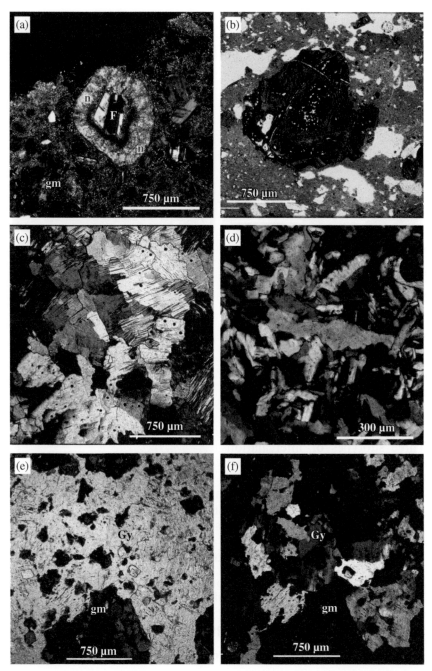

Plate 8.16. *Internal fabric of nodules.* (**a**) nucleic calcite nodule formed around a feldspar (F) grain in groundmass with calcitic crystallitic b-fabric (PPL); (**b**) alteromorph nodule of goethite after garnet; compare also Plate 6.3c and d (PPL); (**c**) xenotopic fabric of gypsum in petrogypsic horizon; notice polycyclic twinning due to dehydration (XPL); (**d**) hypidiotopic fabric of gypsum in petrogypsic horizon; the lenticular shape of the crystals is still visible (XPL); (**e**) poikilotopic fabric: the gypsum crystals (Gy) enclose fragments of the groundmass (gm) (PPL); (**f**) same (XPL).

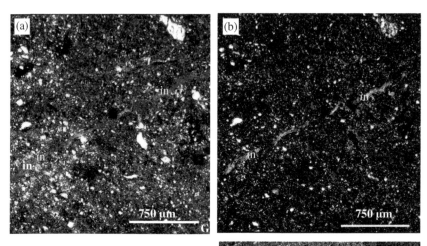

Plate 8.17. *Intercallations*. (**a**) simple intercalation (in) of clay (PPL); (**b**) same, notice internal parallel orientation of the clay (XPL) (**c**) irregular intercalation (ic) and (in) of neoformed glauconite in saprolite near its contact with sea water (PPL).

Intercalations are not always clearly distinguishable from coatings of planar void that are now closed, from which the original voids disappeared, nor from planar void infillings.

8.7.3 Classification

Three subtypes of intercalations are distinguished based on their external shape (Fig. 8.15):

Simple: individual intercalations (Plate 8.17a through c);
Serrated: intercalations with serrated ends;
Interlaced: intercalations consisting of interwoven strands.

8.7.4 Descriptive Criteria

Description of an intercalation comprises its nature, internal fabric, and especially its referred or related orientation pattern.

8.8 EXCREMENTS

8.8.1 Introduction

Excrements take a very special position among the pedofeatures, because they can belong to different levels since they can occur on the level of microstructure (e.g., granular and intergrain microaggregate microstructure), groundmass (e.g., in vermic materials), fabric pedofeatures (e.g., passage features), and material forming infillings or even coatings. It was therefore not possible to incorporate them in a key nor to establish a significant

a   b   c

Fig. 8.15. Types of intercalations (a) simple, (b) serrated, and (c) interlaced. (From of Bullock et al. 1985.)

classification. Excrements of soil mesofauna could be expected to be very important features, yielding information on the types of animals present, the same way hunters look at droppings when following the tracks of their prey. This is generally impossible because the shape and fabric of excrements depends on the age of the animal, its diet, and environment. Several authors (e.g., Babel and Vogel, 1989) demonstrated that it is very difficult, and usually impossible, to link the morphological aspect of excrements of the mesofauna to the kind of animal that produced them.

In Bullock et al. (1985), excrements were in fact limited to droppings of the soil mesofauna. Investigations of archaeological sites have since drawn attention to excrements of larger animals (dogs, sheep, and cattle) which form an important source of environmental information (Courty et al., 1994; Brönnimann et al. (2017a, b). As they are of a totally different order, the small excrements of the soil mesofauna are discussed in Section 8.8.2 and characteristics of some larger excrements in Section 8.8.3.

8.8.2 Descriptive Criteria for Excrements of the Mesofauna

8.8.2.1. Shape

Shape is the most characteristic feature of excrements. Several types (Fig. 8.16, Table 8.4) were distinguished by Bullock et al. (1985) including

	LONGITUDINAL SECTION	CROSS SECTION		LONGITUDINAL SECTION	CROSS SECTION
spheres	⬭	⬭	polled bacillo-cylinders	⬭	⬭
ellipisoids	⬭	⬭	bipointed cylinders	⬭	⬭
conoids	⬭	⬭	clonocylinders	⬭	⬭
			grooved plates	⬭	⬭
tailed conoids	⬭	⬭	mitoids	⬭	⬭
pointed tailed conoids	⬭	⬭	mammilated exrements	⬭	⬭
cylinders	⬭	⬭	tuberose exrements	⬭	⬭
bacillo-cylinders	⬭	⬭			

Fig. 8.16. Types of excrements. See also table 8.4 (from Bullock et al. 1985).

Table 8.4. The main shapes of intact excrements of the soil mesofauna (after Bullock et al. 1985) (see also Fig. 8.16)

Type	Description	Possible origin
Spheres	The excrements have a spheroidal habit, i.e., they always have the shape of a circle in thin section.	Larvae of Adelidae and Bibionidae
Ellipsoids	The excrements are egg-shaped, i.e., they have in longitudinal section the shape of an ellipse, and in cross-section that of a circle.	Orbatid mites
Conoids	Cone-shaped excrements. In thin section the longitudinal section is an "asymmetrical" ellipse much broader on one side than the other; in cross-section the excrement is circular.	Julidae and Glomeridae
Tailed conoids	Conoids of which the broadest side ends in a small sharp point.	Julidae and Glomeridae
Pointed tailed conoids	Like tailed conoids, but the smallest side has a pointed end.	Julidae and Glomeridae
Cylinders	The excrements have the shape of a cylinder with flat ends.	Larvae of Lymnophilidae
Bacillo-cylinders	The excrements have the shape of a cylinder with rounded ends.	Enchytraeids, some Orbatid mites; several Lumbricidae
Polled bacillo-cylinders	The excrements have a cylindrical shape; one end is rounded, the other is flat.	Larvae of Tipulidae.
Bipointed cylinders	The excrements are cylindrical; both ends are cone-shaped.	Hydrobia ulva
Clonocylinders	The excrements have a cylindrical shape with some constrictions.	Larvae of several Adelidae and Bibionidae
Grooved plates	The excrements are plates which show a length-wise groove on one side.	Isopodes
Mitoids	The excrements are long threads, usually rolled up.	Helicidae and several Arionidae
Mammillated excrements	Bodies with mutually interfingering spheroidal surfaces.	Lumbricidae
Tuberose excrements	Bodies with irregular rounded protuberances	Lumbricidae

spheres (Plate 4.5b, Plate 8.18a, Plate 8.19a and b), *ellipsoids* (Plate 8.18b and c), *conoids*, *cylinders* (Plate 8.18d), *mammillated*, and *tuberose* excrements. The above mentioned shapes are three-dimensional. As excrements mostly occur as populations, it is usually easy to reconstruct the three-dimensional shape from the two-dimensional observation in thin sections.

8.8.2.2. Aging

As a result of aging, excrements, especially those rich in organic matter, lose their original shape. Three types of aging (alteration) were distinguished by Bullock et al. (1985): disintegration, coalescence, and internal aging. As the latter two are morphologically similar, they are grouped here under the term coalescence.

   1. Disintegration: characterized by the formation of cracks, followed by a loss of shape. This is particularly the case of excrements composed of loosely packed tissue fragments. Three grades of disintegration

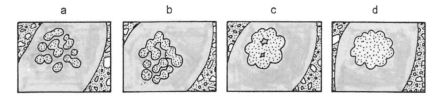

Fig. 8.17. Grades of excrement coalescence and micro-aggregates: (a) very porous, (b) porous microaggregates, (c) dense, and (d) very dense.

are distinguished, based on the volume affected: weak (< 30%), moderate (30–70%) and strong (> 70%).

2. Coalescence: the excrements, or fragments thereof, coalesce at their points of contact and form higher units (microaggregates). Four grades of microaggregates are identified by Bullock et al. (1985) (Fig. 8.17) (Plate 8.19d):

2.1 Very porous microaggregates: a cluster of excrements only linked at the points of contact. The original shapes of the excrements are rather well preserved, and porosity is characterized by a dominance of compound packing voids (Plate 8.19c).

2.2 Porous microaggregates: the original shapes are less clear due to a more generalized coalescence of the excrements or their fragments; a moderate amount of compound packing voids is present (Plate 8.19e).

2.3 Dense microaggregates: original shapes have practically disappeared, and the porosity is characterized by small star-shaped vughs (Plate 8.19d).

2.4 Very dense microaggregates: only at the edge of the aggregate can shapes of the original excrements be recognized; porosity is lacking.

8.8.2.3. Other Criteria

Roughness is related to the composition of the excrements (type of food consumed), the bite sizes of the soil animals and the degree of decomposition. For example, fermentation may lead to rough boundaries.

Size of excrements is linked to the type of animal, but can vary a great deal, depending on its age and diet. Sorting of excrements is mostly very good. Sizes should be indicated as explained in Section 4.4.2.

Color and limpidity can give some information on composition and the degree of humification. Fresh organic excrements generally have vivid yellowish to reddish colors (Plate 8.18b and c), whereas older ones become darker and less limpid and their boundaries become less regular (Plate 4.5b, Plate 8.19c through e).

Composition is one of the most important facts to be recorded. Most excrements are a mixture of organic and mineral matter, but pure organic ones are also common (e.g., in moder), whereas dominantly inorganic ones are rare (Plate 8.19a and b). Depending on the type of animal and diet (e.g., difference between primary and secondary consumers) the organic fraction may be composed of tissue fragments, cells and cell residues (e.g., fragments of fungal hyphae) (Plate 5.2b, Plate 8.19e), fine organic matter or a mixture. Ultraviolet fluorescence is a useful tool for analyzing excrements.

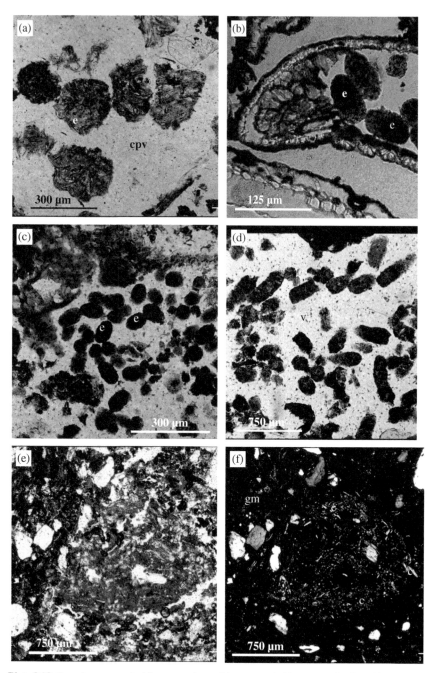

Plate 8.18. *Organic excrements.* (**a**) coarse spheres (e) composed of fragments of cells and fungal hyphae; compound packing voids (cpv) (PPL); (**b**) organic ellipsoids (e) (orbatids) in plant fragment (PPL); (**c**) fresh organic ellipsoids (e) (mites) (PPL); (**d**) organic cylinders (PPL); (**e**) herbivore dropping (central part, lighter area) composed of organic fine material and cell fragments; for detail see Plate 6.8a and b (PPL); (**f**) same, notice interference colors of cellulose (c) and calcite spherulites; undifferentiated b-fabric in groundmass (gm) (XPL).

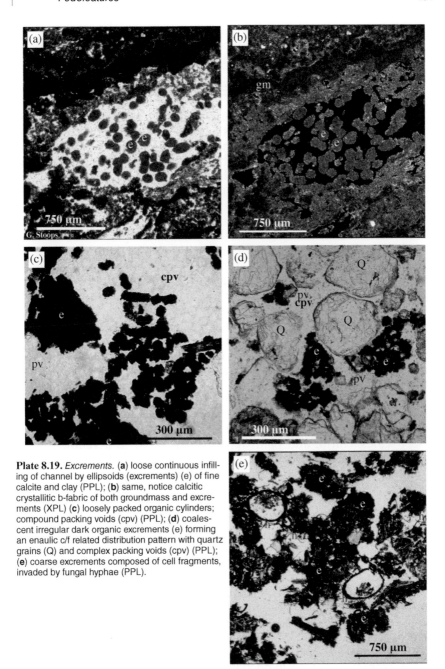

Plate 8.19. *Excrements*. (a) loose continuous infilling of channel by ellipsoids (excrements) (e) of fine
calcite and clay (PPL); (b) same, notice calcitic
crystallitic b-fabric of both groundmass and excrements (XPL) (c) loosely packed organic cylinders;
compound packing voids (cpv) (PPL); (d) coales-
cent irregular dark organic excrements (e) forming
an enaulic c/f related distribution pattern with quartz
grains (Q) and complex packing voids (cpv) (PPL);
(e) coarse excrements composed of cell fragments,
invaded by fungal hyphae (PPL).

Internal fabric comprises basic and referred orientation and distribution patterns (e.g., concentric, crescent like) and the c/f-related distribution, which can be for instance monic (in the case of tissue fragments) or
porphyric.

BACKGROUND – Brewer (1964a) used the term fecal (or fecal pellet) to describe excrements. Two types are distinguished: pellets and welded pellets; no further subdivisions were proposed.

8.8.3 Descriptive Criteria for Excrements of Larger Animals

Shape, composition, and internal fabric of the most important large excrements found in archaeological contexts (mainly caves and shelters) are discussed by Courty et al. (1994), Brönnimann (2017 a and b) and Karkanas (2017). External shape and microfabric are important characteristics to identify excrements of larger animals. As their size exceeds that of most thin sections, only some general properties will be mentioned below. For more details, the reader is referred to the papers mentioned before.

Carnivore excrements: bright yellow, containing well-crystallized phosphatic material and bone fragments; they show a strong UVF and BLF.

Omnivore excrements: their composition depends on the type of food digested. Excrements of omnivores on diet rich in biological products show characteristics similar to those of carnivores, whereas a vegetarian diet produces excrements bearing characteristics of herbivores. Combinations of both are common in natural conditions.

Herbivore excrements: can be recognized by the presence of fragmented plant material, often with large amounts of phytoliths, Ca-oxalate crystals, calcite spherulites (Plate 6.8a and b, Plate 18e and f) and sometimes diatoms. They have a spongy fabric that may become denser with age.

Birds and bats excrements: have a spongy microstructure when fresh, altering to a denser one with time. The composition depends on the feeding habit of the animals. Partly dissolved fine bone fragments, insect scales, chitin, and plant remains can occur in a yellow micromass. Most guano shows UVF or BLF (Karkanas, 2017).

8.9. COMPOUND PEDOFEATURES

Compound pedofeatures are those that consist of a mixture of two or more pedofeatures. They point either to a change in pedogenic processes, or to the simultaneous action of different processes, and are as such important records of environmental conditions. Two types of compound pedofeatures can be distinguished (Fig. 8.18):

Juxtaposed: the different units lie side by side, for example, a calcite coating covering a clay coating, a goethite coating covering an Fe hydroxide hypocoating (Plate 3.4e and f; Plate 6.5e and f; Plate 8.1a through d; Plate 8.7a and b, Plate 8.8e and f; Plate 8.9f; Plate 8.20; Plate 8.21a and b).

Superimposed: the different units are superimposed to one another, for example a calcite hypocoating developed in a clay coating, Fe nodules developed in clay coatings.

The terminology described above applies to an individual feature such as a compound coating. Fedoroff and Courty (1994) discussed the related distribution for a population of compound pedofeatures, distinguishing two types (A and B): *juxtaposed*, where A always post-dates B (Plate 6.5e and f; Plate 8.20c through d), and *imbricated*, where

A and B occur in random order. This is especially useful in the description of clay coatings

In classifying and describing compound pedofeatures, each fabric unit is treated separately, that is, as simple pedofeatures, and later recombined using the concepts of juxtaposition and superimposition.

8.10. COMPLEX PEDOFEATURES

In most cases the juxtaposition or superimposition happens by chance, as a result of changing local or environmental conditions. However, some combinations are observed more frequently, and a genetic relationship between the individual features is supposed or demonstrated. Such associations of pedofeatures will be called complex pedofeatures (Stoops, 2003). They have a specific diagnostic value. A few examples have been described in literature, some even have specific names. As much more research is needed on this topic, only a few examples will be given below.

In semiarid calcareous soils, channel infillings of coarse (cytomorphic) calcite crystals, pseudomorphic after root tissues, are frequently surrounded by a calcite depletion hypo-coating. This combination has been described in detail by Herrero and Porta (1987) and Jaillard et al. (1991). Herrero and Porta (1987) proposed a specific terminology: the complex feature as a whole is called a *quera*; it consists of a channel filled with *quesparite* (consisting of calcite phytoliths or cytomorphic calcite crystals),

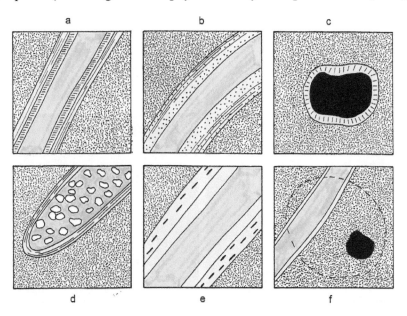

Fig. 8.18. Superimposed, juxtaposed and imbricated pedofeatures: (a) juxtaposed: typic coatings, (b) superimposed: ferruginous silt coating (dotted with streaks) with iron-depletion hypo-coating (dotted only), (c) juxtaposed: Mn-hydroxide nodule with calcite coating, (d) juxtaposed: channel with typic coating and loose continuous infilling, (e) juxtaposed and superimposed: two juxtaposed clay coatings; Fe-hydroxide micronodules superimposed to the first one, and (f) superimposed: weak impregnation of Fe-hydroxide (outlines indicated by discontinuous circle) superimposed to clay coating and Mn-hydroxide nodule.

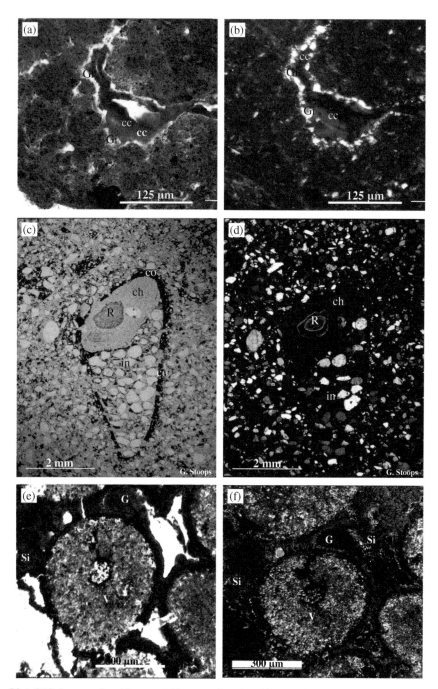

Plate 8.20. *Compound pedofeatures.* (**a**) juxtaposed coating of (colorless) gibbsite (Gi) and reddish fine clay (cc) (PPL); (**b**) same; gibbsite is clearly visible (XPL); (**c**) coating of organic matter (co) on channel (ch) wall formed by termites and loose partial infilling by sand grains (in); in the upper part of the channel a root section (R) is visible (PPL); (**d**) same, notice interference colors of the root remains (XPL); (**e**) root channels (root residues still visible in center) filled with fine grained blue vivianite (V) and surrounded by a dark brown siderite (Si) coating; fibrous goethite (G) in the interstitial space (PPL); (**f**) same; isotropic goethite (G), root, siderite (Si) and vivianite (V) (XPL).

surrounded by the calcite depleted hypo-coating (the *quedecal*) and with associated perpendicular microchannels (the *quevoids*) (Plate 8.21c–f).

Also the *microband fabrics* observed in seasonally frozen soils (Dumanski and St.Arnaud, 1966; Mermut and St. Arnaud, 1981; Van Vliet-Lanoë et al., 1984; Van Vliet-Lanoë and Fox, 2018) are another example of complex pedofeatures, characterized by a typic lenticular platy microstructure, with a concentration of finer particles in the upper part of the peds and of bare coarse particles in the lower part, or with a textural gradient (Plate 8.22a and b).

8.11 FRAGMENTED, DISSOLVED AND DEFORMED PEDOFEATURES

Due to processes such as pedoturbation, bioturbation, and dissolution pedofeatures can be fragmented, dissolved and/or deformed, and should be described as such. Changes resulting from chemical or mineralogical alterations that do not influence the morphology of the features are not considered here. During pedo- and bioturbation, rigid pedofeatures (e.g., calcite or Fe nodules) are often fragmented, whereas less rigid pedofeatures (e.g., most clay coatings or infillings) are easily deformed. These changes depend on the type of forces exerted, and the humidity of the soil at the moment of deformation, especially in the case of swelling clays. For example, animal activity can fragment clay coatings, shear stress will deform them; clay coatings will be fragmented in a dry state and deformed when moist.

Fragmented pedofeatures (Fig. 8.19) may have one or more of the following characteristics: (i) a discontinuity of the internal fabric at the boundary and (ii) sharp boundaries at the place of fragmentation, often with sharp edges. If not recognized as such, fragments of pedofeatures will automatically key out as nodules (Plate 8.22c through e).

> BACKGROUND – In Brewer (1964a) fragments of layered clay or silt coatings were classified as papules (subdivision of glaebules). The use of this term is strongly discouraged, as also weathered biotite and shale fragments are covered by that term.

Deformed pedofeatures. Deformation is generally observed in clayey pedofeatures. They may show the following characteristics: (i) an elongate, frequently undulating form, (ii) diffuse boundaries, and (iii) a deformed internal fabric (Fig. 8.20, Plate 8.23a and b).

In the case of clay coatings, the deformation of the internal fabric is generally expressed by a change of the continuous orientation of the clay to a striated one. A transition to a kink-band fabric is only rarely observed (Plate 8.22f; Plate 8.23c and d).

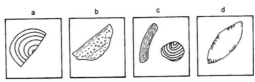

Fig. 8.19. Examples of fragmented and partly dissolved pedofeatures: (a) fragmented concentric nodule, (b) fragmented surface crust, (c) fragmented layered clay coatings and infillings, and (d) partly dissolved lenticular gypsum.

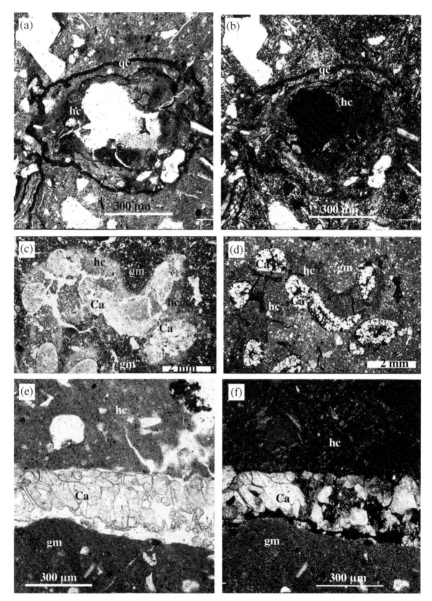

Plate 8.21. *Compound pedofeatures*. (**a**) juxtaposed pedofeatures; channel with coating of coarse illuvial clay, Fe- hydroxide hypo-coating (hc) and quasi-coating (qc) (PPL); (**b**) same, notice strong developed striated b-fabric in groundmass (bottom) (XPL); Complex pedofeatures. (**c**) channel in calcareous groundmass (gm) surrounded by calcite depletion hypo-coating (lighter area), and with loose discontinuous infilling of sparitic (cytomorphic) calcite (Ca); the whole feature is known as a "quera" (PPL); (**d**) same, notice calcitic crystallitic b-fabric of groundmass (gm) and the speckled b-fabric of the Fe-depletion hypocoating (hc) (XPL); (**e**) channel with calcite depletion hypo-coating (hc) (top) and dense incomplete cytomorphic sparite (Ca) infilling; the lower part represents the normal groundmass (gm); "quera" (PPL); (**f**) same, notice calcitic crystallitic b-fabric in the groundmass (gm) and striated b-fabric in decalcified hypocoating (hc) (XPL).

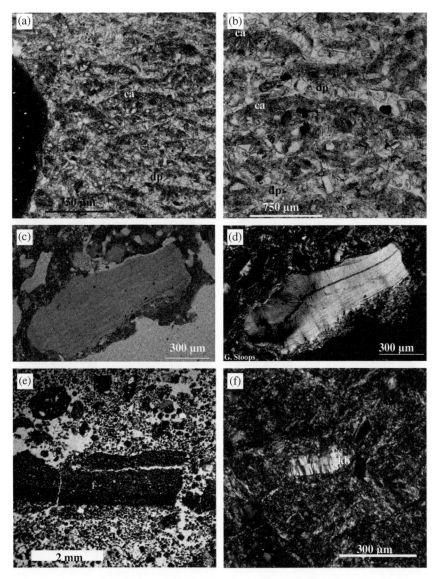

Plate 8.22 *Complex pedofeatures.* (**a**) microband fabric, consisting of lenticular aggregates with capping (ca) of fine material on top, and depletion (dp) of fine material at their base, due to frost action; at left part of an anorthic Fe-Mn-hydroxide nodule (PPL); (**b**) detail (PPL); *Fragmented pedofeatures.* (**c**) subangular fragment of microlaminated clay coating (PPL); (**d**) same (XPL); (**e**) fragment of surface crust (PPL); (**f**) small coating fragment (cc) with clear kink band (kb) deformation (XPL).

Dissolved pedofeatures are the result of congruent dissolution, resulting in the appearance of indentations not compatible with the internal fabric of the features. The shape of the indentations depends on the nature of the feature. In the case of coarse crystalline pedofeatures (such as crystals, crystal intergrowths, nodules and coatings) dissolution often follows crystallographic zones of weakness, resulting in serrated outlines. Congruent dissolution of microcrystalline or amorphous pedofeatures (e.g., Fe or Mn (hydr)oxide nodules) often produces irregular embayments.

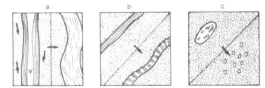

Fig. 8.22. Examples of deformed pedofeatures: (a) illuvial clay coating on a plane (V) (left) is deformed by pressure and shearing (right); due to pressure the central plane (V) is closed; shearing results in a loss of continuous orientation of the clay particles, and the appearance of a striated one; (b) fragment of an illuvial clay coating (left) is deformed by pressure (right); the external form has changed, and a kink-band fabric is superposed to the internal parallel clay orientation; (c) loose discontinuous channel infilling with lenticular gypsum (left) is destroyed by bioturbation, resulting in random clusters of isolated gypsum crystals in the groundmass (right).

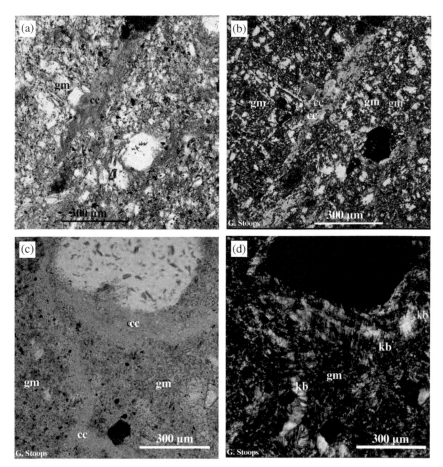

Plate 8.23 *Deformed pedofeatures* (**a**) stress deformed clay illuviation coating (cc), incorporated in groundmass (PPL); (**b**) same, notice loss of continuous orientation in coating (no extinction bands), ranging to a striated one (XPL); (**c**) fragment of limpid yellowish clay coatings (cc) incorporated in speckled groundmass (gm); notice abrasive powder in channel (PPL): (**d**) same, notice kink band (kb) deformation in coatings (XPL).

9. Making and Presenting Thin Section Description

9.1 INTRODUCTION

A thin section description is the final result of fabric units identification processes. It should not be considered as an isolated study, but rather as an integrated research tool used in interaction with other techniques. Indeed, the more information that is available on the environment and the field aspects of the soil or regolith profile, such as the chemical, physical, and mineralogical composition of the layer from which the sample was taken, then the better the micromorphologist can direct her or his observations and choice of methods to be used. This starts right from the beginning when the selection of drying and impregnation methods must be made based on the field conditions at the time of sampling, type of material and aim of the study. Even in the study of isolated samples of earthy materials (e.g., nests of insects, mortars, plasters or contents of jars) it is necessary to have as much information as possible on the sample's position and environment. Micromorphological descriptions can help colleagues choosing the most adequate chemical, mineralogical and physical analyses to be made.

As far as possible the micromorphological study should be performed by the scientist that studied the profile and took the samples, but under the guidance of an experienced micromorphologist. This is especially true for the interpretation. Only few soil scientists have a sufficient knowledge of the broad international literature in this field (see Section 2.3.4). Moreover, it is necessary to have at least a basic knowledge of optical crystallography, mineralogy, and petrography to be able to determine components adequately. This knowledge is often missing among young researchers, as formulated by Shahack-Gross (2015), as geology and mineralogy got a lower priority in education last decennia.

In some cases, however, the micromorphologist may also be asked to make a description of and form an opinion on a material without any additional information to avoid being biased by additional data. This generally

Guidelines for Analysis and Description of Soil and Regolith Thin Sections, Second Edition. Georges Stoops.
© 2021 Soil Science Society of America, Inc. Published 2021 by John Wiley & Sons, Inc.
doi:10.2136/guidelinesforanalysis2

happens when other scientists have a scientific dispute, but such situations are fortunately rare. Such "blind description" may be proposed to students as a puzzle: to guess what their object is, or in a research to make it less dependent on the conceptual background of the micromorphologist.

9.2 OBSERVATION

"In front of a piece of art one should not dream, but one should make an effort to understand its meaning. The more one observes, the more mysterious the simple matter becomes. We see, so we think, clearer when we analyse, when we look to every detail, but when pieces have been taken apart we are generally not able to bring them together, and lost what we possessed already, the global with all its details and soul." (Arnold Schönberg, Austrian/American composer, 1874/1951).

The same can be said regarding thin section studies. Among laymen and beginner micromorphologists, there is generally a strong belief that the higher the magnification used, the better the features of the studied material can be seen. This is often not the case. Two different soil materials may be quite easily distinguished with the naked eye, but become indistinguishable at high magnification. Analogous to this are landscape features, only revealed on aerial photographs or satellite images, but invisible on the ground. It is therefore recommended to start every thin section study at the lowest magnification possible (e.g., using objective 1x). A first inspection of the section with the naked eye or a binocular microscope may reveal a lot of information on larger fabrics, structures, and heterogeneities not visible at higher magnifications. Each feature should then be studied gradually at increasing magnifications.

During the inspection of the thin section at this low magnification, specific large features can be marked on the coverglass of the section with a soft-tipped marker. It is often also useful to mark, with a dashed line, at this stage the boundary between homogeneous zones in the case of a mesoheterogeneous material. This can also be done on a printed scan of the thin section. This first mesoscopic observation should also allow the researcher to judge the quality of the section.

Before starting the microscopic observations, the thin sections, especially their cover glasses, should be cleaned carefully with tissue paper. Dust or fingerprints on the cover glass will result in more diffuse images, especially at higher magnifications, resulting in an excessive tiring of the eyes and blurring micrographs. Uncovered thin sections have to be covered with glycerine or a special oil to enhance contrast and to deduce effects of lapping. Objective lenses and eyepieces of the microscope should be kept clean using only lens-tissue, or microfiber tissue.

Next, observe the fabric of each homogeneous zone with increasing magnifications. Be aware that the image seen in the microscope is upside down. Descriptions are made essentially in plane-polarized light (PPL), supplemented with cross-polarized light (XPL). For sandy materials [especially with monic, enaulic and chitonic c/f (coarse/fine) related distribution patterns] the use of partly-crossed polarizers, or 1/4 λ retardation plate (see Section 3.2.2.1) might be useful, as this summarizes the complementary images observed in PPL and XPL. Whenever necessary, studies should be completed with observations in oblique incident light (OIL), UV, CL, etc.

When possible, it is useful to combine thin section studies with mesomorphological studies of undisturbed, non-impregnated samples of the same material, using a stereomicroscope. This may help a better understanding of the three-dimensional spatial arrangement, and allowing to separate by hand-picking specific constituents or features for additional analyses (e.g., chemical, XRD, WDS).

9.3 ARTIFACTS

Some of the most common defects and failures in thin sections that can interfere with observations and descriptions are summarized in Table 9.1. Examples of possible confusions are crystallizations in voids that could be described as crystal intergrowths, and abrasive powder in the micromass, that give the impression of organic microparticles.

9.4 DESCRIPTION

9.4.1. How to Start?

When studying complete soil or regolith profiles, it is worthwhile to have a look at all thin sections before starting the individual descriptions. This way of working may save a lot of time because features, difficult to recognize or

Table 9.1. Common defects in thin sections.

Artifact characteristics	Cause	Plates
Sharp circular or somewhat irregular, mammillated features, colorless, with very high relief, in voids and in focus together with the soil material	Air bubbles in resin	Plate 9.1d
Sharp circular or sometimes irregular features, colorless, with very high relief, not in focus together with the soil material	Air bubbles in mounting medium of support- or cover glass	Plate 9.1a through c
Parallel streaks with lighter colors in the micromass (PPL) and with lower interference colors (XPL) in minerals	Grooves made by the grinding wheel of a plane polishing machine	
Zones with diffuse whitish to gray interference colors and often with wavy extinction in pores (XPL)	Stress birefringence caused by strain in resin	Plate 9.1d and e
Misty zones over thin section, generally more grayish in PPL	Probably reaction of mounting medium with oil used during polishing	Plate 9.2a
Small crystals or flower like crystal aggregates with interference colors of first to second order, sometimes restricted to voids and simulating coatings, or superposed to soil constituents (XPL)	Crystallization (devitrification) of resin or mounting medium, especially in older thin sections	Plate 9.2b through d
Small angular grains with very high relief occurring both in the fine material and voids	Grains of abrasive powder	Plate 6.7c, Plate 8.23a, Plate 9.1d
Polysynthetic twinning of gypsum (XPL)	Dehydration as a result of heating (e.g., when using Canada balsam for mounting the section)	Plate 8.16 c, Plate 9.2e
Zones with higher and lower interference colors for the same mineral (XPL)	Uneven thickness of the section, parts > 30 µm thick	Plate 9.2f

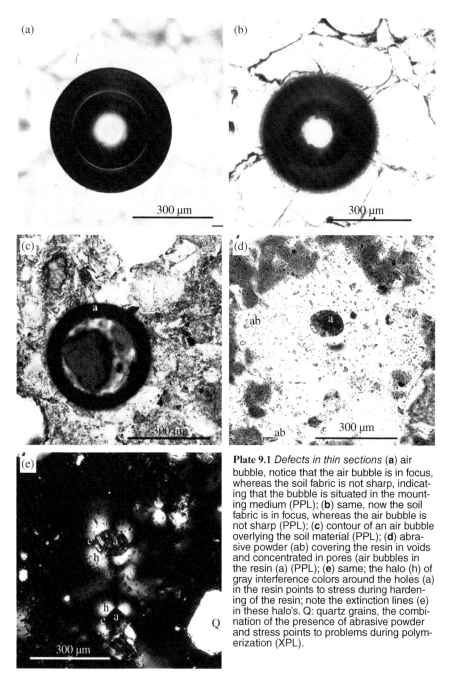

Plate 9.1 *Defects in thin sections* (**a**) air bubble, notice that the air bubble is in focus, whereas the soil fabric is not sharp, indicating that the bubble is situated in the mounting medium (PPL); (**b**) same, now the soil fabric is in focus, whereas the air bubble is not sharp (PPL); (**c**) contour of an air bubble overlying the soil material (PPL); (**d**) abrasive powder (ab) covering the resin in voids and concentrated in pores (air bubbles in the resin (a) (PPL); (**e**) same; the halo (h) of gray interference colors around the holes (a) in the resin points to stress during hardening of the resin; note the extinction lines (e) in these halo's. Q: quartz grains, the combination of the presence of abrasive powder and stress points to problems during polymerization (XPL).

rare in one layer, may be very clear and/or common in another one. It also gives a general impression on the study object for better understanding its individuality and for adequate approach to its ingredients (not to exaggerate some features). Moreover, it makes identification of interhorizon transport (e.g., in infillings) clear right from the first section described.

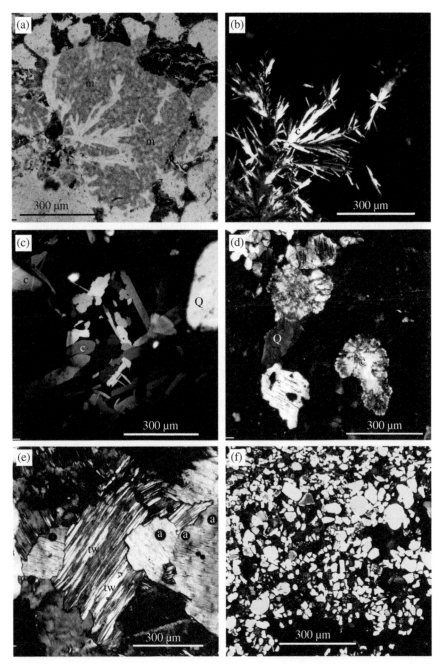

Plate 9.2 *Crystallizations in the resin* (**a**) misty (m) area in mounting medium (PPL); (**b**) same, crystallization of resin in mounting medium (XPL); (**c**) crystallization (c) of the resin used to mount the cover glass; notice that some of the crystals overlap with quartz grains (Q) (XPL); (**d**) crystallization (c) of the resin used to mount the section; notice that the artificial crystal aggregates (with high interference colors) overlap with underlying quartz grains (Q) (XPL); (**e**) polysynthetic twinning (tw) caused by dehydration of gypsum as a result of heating in thin section mounted with Canada Balsam; circular features are air bubbles in the mounting medium (a) (XPL); (**f**) quartz sand; the interference colors of some grains reach often first order red, indicating that the section is locally too thick (> 30 µm) (XPL).

It is recommended to start at the lowest magnification, describing the largest fabric units first, then analyzing the simplest units of which they are composed, and then ending with the basic components.

This approach can be considered as a natural continuation of the field observations, which can be continued on the submicroscopic scale. It also allows the analyst to stop at a level where accumulating more information is no longer required for the aim of the study. It is impossible, and often useless, to describe all details of a thin section. The degree of detail to be given depends on the aim of the description.

Descriptions should always be made in relation to the study's objective. For instance, studies made to support research on soil physics will generally not require a detailed analysis of the origin of the coarse material, or even of its mineralogical composition, whereas these data are very important for pedogenic studies. Patterns of porosity, of primary interest in soil physics, may be less important in archaeological research. In reality, every grain, every coating, or every nodule is unique; describing each individually, even if possible, would be far from efficient. In view of this it is clear that the scientist has to describe populations of grains or of features (partial fabrics), indicating the variability within the group of fabric units.

Another possible approach would be to start at high magnification (e.g., objective 25 or 40) with the identification and description of the simplest units (e.g., mineral grains, micromass). The analyst studies how these units are combined into larger units with increasing degree of complexity, arriving finally at the macroscopic scale. This means that one tries to synthesize the complex soil fabric starting from the simplest building blocks. The drawback is that continuity with the mesomorhological obervations and the field are lost, and that probably much time is lost in determing features not important for the aim of the research.

In temperate soils on uniform sedimentary material, such as eolean sands or loess, where pedogenic processes act mainly downward, most micromorphologists start with the study of the upper horizons, and gradually move to the deeper horizons, ending finally in the parent material. In deep tropical weathering profiles, it is generally easier to start with the deepest material (generally a coherent rock) and successively move to the higher layers or horizons, as this has the advantage that the scientist can gradually observe the neoformations (e.g., alteromorphs) difficult to identify otherwise.

In thin section descriptions it is recommended that the expression "not detected" or "not observed" be used instead of "absent" for features such as coatings or small crystals. The detection limit of optical microscopes does not allow the observation of features below 10 to 20 μm, unless they are sufficiently contrasting (see also Section 3.1.4). Remember that the absence of evidence is no evidence of absence. Moreover, a thin section is not necessary representative for the whole horizon, especially in heterogeneous materials.

A description system is a convention. It is therefore absurd to use a mixture of two or more systems. In each system agreements are made as to what the user and the reader should understand by each concept; a mixture of systems thus often results in contradictions. One can best compare it with traffic regulations: it is not possible to drive sometimes at the right side, sometimes at the left side of the road depending on the landscape one wants to admire.

Therefore, only one system of terminology should be used in a description. For example, "an s-matrix with porphyric c/f related distribution and masepic plasmic fabric" makes no sense, as "s-matrix" and "masepic" are based on the concept of plasma and skeleton grains (Brewer, 1964a), whereas the c/f-related distribution is based on the idea of coarse and fine material, which do not necessarily have the same characteristics and limits as plasma and skeleton grains. The terminology recommended in several languages may be found in Appendix II, or at http://wwwisric.org/explore/ISRIC-collection/micromulti

9.4.2. Possible Description Schemes

For a general standard description the author recommends following scheme:

Microstructure and Porosity Pattern

- Type of microstructure(s), size and shape of aggregates, optionally with relative proportions and patterns if several types occur; hierarchy of microstructure. For sandy materials the basic microstructure, corresponds generally to the c/f-related distribution pattern, when the latter is not porphyric.

- Intra-and transpedal voids, their abundance, size, and pattern. Packing voids are generally not mentioned as they are inherent to basic microstructures.

Groundmass

- *c/f limit, c/f ratio and c/f related distribution,* if different from the basic microstructure.

- **Coarse material (mineral and organic separately):** nature (including weathering or humification), size, frequency, and for each population the orientation and distribution pattern when not at random.

- **Micromass:** nature (if it can be determined), color, limpidity, interference colors (if any), and b-fabric.

Organic material not included in the groundmass or in pedofeatures (e.g., in channels or root sections):

- type
- degree of alteration
- size
- pattern.

Pedofeatures

A grouping of pedofeatures according to their composition seems appropriate, for example, calcitic features, clayey features, and Fe-oxide features. A grouping according to processes involved (e.g., illuviation, oxido-reduction) is generally excluded because it already supposes an interpretation, and is thus in contradiction with the pure morphological basis of the system. Grouping according to morphological types is not recommended (e.g., creating a group of all coatings that would combine clay and calcite coatings under a same heading, whereas clay and calcite infillings, which might be aspects of a same process, would appear under a different heading). Nature, size, shape, variability, abundance, and pattern of each population of pedofeatures must be noted.

9.4.3 Examples

Examples of short systematic descriptions, as they were made at the International Training Centre for Post-Graduate Soil Scientists of the Ghent University, are given below.

Example 1

Profile N° xx, horizon: Bt, sample n°: xxxx, thin section n° xxxx

Vertical thin section; size about 40 cm²; thickness: about 20 μm.

Abbreviations: c.s., coarse sand size (2000 μm–500 μm), m.s., medium sand size (500–125 μm), f.s., fine sand size, (125–50 μm), s.s., silt size (< 50 μm).

Microstructure and Porosity

Moderately separated and well developed subangular blocky microstructure with partially accommodating peds (5–10 mm); vughy intrapedal microstructure; mammilated vughs (c.s., ± 5% of the ped); smooth channels (m.s., ± 5% of the ped).

Groundmass

Clf*$_{5μm}$ *ratio: 1/4; c/f related distribution pattern: double spaced porphyric.

Coarse Material:

Mineral: angular and subangular grains of quartz (m.s.), dominant; subangular grains of plagioclase (m.s.), ± 10% of the coarse fraction; angular flakes of biotite and muscovite (c.s.), ± 5%; quartz aggregates (m.s.), ± 5%, rounded grains of epidote (f.s.) ± 2% with irregular linear weathering.

Organic: few dark brown tissue residues (m.s., 2%);

Micromass: yellowish brown, speckled clay with weak cross-granostriated and stipple-speckled b-fabric.

Organic material not included in the groundmass or in pedofeatures: fresh root sections (m.s.) in channels (± 2% of total area).

Pedofeatures

- Microlaminated, continuous coatings of limpid yellowish clay, with lower first order interference colors, about 100 μm thick, on most ped surfaces and channels (± 4% of total area); channel and plane infillings of same material (± 2%); similar deformed fragments in groundmass (± 1%).

- Channel infilled with ellipsoidal yellowish brown speckled isotropic excrements (f.s.).

Example 2

Profile N° xx, horizon: gypsic, sample n°: xxxx, thin section n° 13.556

Vertical thin section; size about 32 cm^2; thickness: about 20 μm.

Microstructure and Porosity

Moderately separated subangular blocky microstructure with undulating, partially accommodating planes; peds (3–4 mm) consist of dense packing of smooth granules (300– 700 μm) and mineral grains. Common smooth channels and vughs (500–3000 μm, 8% of total area).

Groundmass

Clf $_{10μm}$ ***ratio:*** 1/5; c/f related distribution pattern: open spaced porphyric; locally weakly crescentlike basic distribution pattern.

Coarse material:

Mineral: mainly fragments of limestone (micritic or coarse grained, containing forams) (f.s., sometimes c.s.) and subangular grains of calcite (f.s.- s.s), ± 5% subangular grains of quartz, feldspar and mica (f.s. and s.s.).

Micromass: grayish brown speckled mixture of clay and calcite; calcitic-crystallitic b-fabric.

Organic material not included in the groundmass or in pedofeatures: Fresh root residues in channels, with yellowish isotropic cortex, (± 2%); one root contains ellipsoidal, limpid, isotropic, orange excrements of 25 μm diameter.

Pedofeatures

- Loose continuous infillings of larger (> 1.5 mm) vughs and channels with lenticular gypsum crystals (c.s.), locally intergrown, forming massive aggregates of ± 5 mm^2 with hypidiotopic fabric (± 5% of total area);

- Loose continuous infillings of channels (< 1 mm diameter) with lenticular gypsum crystals (s.s.); in one case root remains are still visible; similar infillings in packing pores (± 5% of total area):

- Lenticular gypsum crystals, ranging from s.s. to c.s., in groundmass (10% of groundmass).

9.5 PRESENTATION OF DATA

Complete descriptions are rather lengthy and are as such only suitable to be published in reports, dissertations, or some monographs. For most publications, reporting of micromorphological observations has

Table 9.2. Example of micromorphological description of a profile in a table. Profile MS 2; Aridic Haplustalf; thin sections 24.240 - 24.243. Vertical thin sections, size 60 × 120 mm; thickness about 20 µm.

	Soil depth		
	0 - 10 cm	50 - 60 cm	130 - 160 cm
Microstructure	intergrain microaggregate / spongy with channels	channel (500 µm) and vugh	as above
Groundmass c/f limit: 10 µm			
coarse material mineral	subangular quartz and feldspar (mainly microcline), 7/3 (m.s - f.s) < 3% charcoal	as above, but quartz/ feldspar ratio 6/4	as above
organic	fragments (m.s.)		
micromass	brown speckled clay; undifferentiated b-fabric	lighter brown, speckled clay; weakly speckled b-fabric	as above
c/f related distribution	fine close enaulic / close porphyric	close porphyric	as above
organic material	few root sections;	few root sections	-
Pedofeatures	-	thick (50µm) microlaminated coatings of strongly oriented reddish limpid clay (10 %), partially stress deformed; large (1 cm) continuous infilling by material from overlying horizon	irregular typic moderately impregnated matrix nodule (2 - 3 mm) and hypo-coatings of microcrystalline calcite (5 %); small (500 µm) diffuse, weakly impregnated typic matrix nodules of Mn-hydroxide (< 3 %).

† m.s., medium sand size; f.s., fine sand size

Table 9.3. Comparison between B horizons in soils on basalt and ash of a toposequence on Isla Santa Cruz (Galápagos Islands) (see also Stoops, 2013).

Profile	9	42	3	17
Altitude	470 m	390 m	155 m	20 m
Microstructure†	granular (500 µm) to crumb (2 mm)	granular (75 µm)	ab and sab with channels	ab (500-1000 µm), locally compacted granular (600-700 µm)
Groundmass				
Coarse material‡	mainly fine I (< 100 µm), few F, few large O and I (400 µm)	R and Y isotropic particles (ss)	few subrounded O and I (500-100 µm)	A (100 µm), rounded fresh O, few F, fragment of B
Fine material§ b-fabric	B, dotted undifferentiated	Y, speckled undifferentiated	B, speckled weakly granostriated	RY, speckled grano- and porostriated
c/f related distr.	double spaced porphyric	fine monic	open porphyric	fine monic to open porphyric
Organic matter	root sections, cell residues	much organ, tissue and cell residues	root sections	few root fragments
Pedofeatures¶	n.o.	n.o.	fragments of stress deformed clay-coatings in groundmass	fragments of stress deformed clay coatings; small (150 µm) rounded, sharp Fe-oxide nodules; excrement infillings

† ab, angular blocky; sab, subangular blocky.
‡ A, augite; B, basalt; F, feldspar; I, iddingsite; O, olivine; (ss), sand size
§ B, brown; R, red; Y, yellow.
¶ n.o., not observed

to be restricted to the mention of those features that are of interest for the discussion. The drawback of this policy is that much existing information which could have been of interest for comparative studies by other scientists, is lost.

A useful technique for presentation of micromorphological data of complete profiles or sequences consists of arranging the data in a table, comparing for instance, (i) the horizons of one profile, or (ii) a given horizon (e.g., B-horizons) of different profiles. Examples are given in Table 9.2 and 9.3. In some cases, a brief table with selected micromorphological information can be combined with a sketch of the profile and a set of micrographs.

In the case of a large number of profiles belonging to a catena or sequence, it can be useful to distinguish several facies, grouping samples with almost the same micromorphological characteristics. Examples are given in Stoops et al. (1994) and Stoops (2013).

A general image of the total thin section is often very informative, especially in the case of heterogeneous materials, or when clear horizon boundaries occur. They can be obtained in a very simple way by scanning the thin section as such (PPL), or between two sheets of polaroid (XPL). More sophisticated methods comprise making a mosaic of partly overlapping micrographs, using a special software (Carpentier and Vandermeulen, 2016; Gutierrez-Castorena et al., 2018; Gutiérrez-Rodriguez, 2018). Captions of scans and micrographs should always contain following information: feature(s) illustrated, profile and horizon, magnification (by preference a bar scale), mode used (e.g., PPL, XPL, OIL).

References

Adams, A.E., and W.S. MacKenzie. 1998. A colour atlas of carbonate sediments and rocks under the microscope. Manson Publ., London. doi:10.1201/9781840765403

Adams, A.E., W.S. MacKenzie, and C. Guilford. 1984. Atlas of sedimentary rocks under the microscope. Longman, Essex, UK.

Adams, J. 1977. Sieve size statistics from grain measurements. J. Geol. 85:209–227. doi:10.1086/628286

Adderley, W.P., C.A. Wilson, I.A. Simpson, and D.A. Davidson. 2018. Anthropogenic features. In: G. Stoops, V. Marcelino, and F. Mees, editors, Interpretation of micromorphological features of soils and regoliths. 2nd ed. Elsevier, Amsterdam. p. 753–777. doi:10.1016/B978-0-444-63522-8.00026-7

Agafonoff, V. 1929. Sur quelques sols rouges et Bienhoa de l'Indochine. Revue de Botanie Appliquée et d'Agriculture Coloniale 9:89–90.

Agafonoff, V. 1936a. Sols types de Tunisie. Annales du Service Botanique et Agronomique de Tunisie 12/13:43–413.

Agafonoff, V. 1936b. Les sols de France au point de vue pédologique. Dunod, Paris.

Algoe, C., G. Stoops, R.E. Vandenberghe, and E. Van Ranst. 2012. Selective dissolution of Fe–Ti oxides— Extractable iron as a criterion for andic properties revisited. Catena 92:49–54. doi:10.1016/j.catena.2011.11.016

Altemüller, H.J. 1962. Gedanken zum aufbau des bodens und seiner begriflichen erfassung. Zeitschr. f. Kult. Tech. 3:323–336.

Altemüller, H.-J. 1964. Die anwendung des phasenkontrastverfahrens bei der untersuchung von bodendünnschliffen. In: A. Jongerius, editor, Soil micromorphology. Elsevier, Amsterdam. p. 371–390.

Altemüller, H.J. 1997. Polarisations- und phasenkontrastmikroskopie mit dünnen bodenschliffen. In: K. Stahr, editor, Mikromorphologische

Guidelines for Analysis and Description of Soil and Regolith Thin Sections, Second Edition. Georges Stoops.
© 2021 Soil Science Society of America, Inc. Published 2021 by John Wiley & Sons, Inc.
doi:10.2136/guidelinesforanalysis2

methoden in der bodemkunde. Hohenheimer Bodenkundliche Hefte 40. Universitat Hohenheim, Stuttgart, Germany. p. 21–88.

Altemüller, H.J., and B. Van Vliet-Lanoë. 1990. Soil thin section fluorescence microscopy. In: L.A. Douglas, editor, Soil micromorphology: A basic and applied science. Elsevier, Amsterdam. p. 565–579.

Altemüller, H.-J., and A. Vorbach. 1987. Veränderung des bodengefüges durch wurzelwachstum von maispflanzen. mitt. Dtsch Bodenkundl. Gesellsch. 55(I):93–98.

Angelini, I., G. Artioli, and C. Nicosia. 2017. Metals and metal-working residues. In: C. Nicosia and G. Stoops, editors, Archaeological soil and sediment micromorphology. John Wiley & Sons, Ltd, Chichester. p. 213–222. doi:10.1002/9781118941065.ch26

Arocena, J.M., and J.D. Ackerman. 1998. Use of statistical tests to describe the basic distribution pattern of iron oxide nodules in soil thin sections. Soil Sci. Soc. Am. J. 62:1346–1350. doi:10.2136/sssaj1998.036159950062 00050029x

Arocena, J.M., G. De Geyter, C. Landuydt, and U. Schwertmann. 1989. Dissolution of soil iron oxides with ammonium oxalate: Comparision between bulk samples and thin sections. Pedologie (Gent) 39:275–298.

Arocena, J.M., G. De Geyter, C. Landuydt, and G. Stoops. 1990. A study on the distribution and extraction of iron (and manganese) in soil thin sections. In: L.A. Douglas, editor, Soil micromorphology: A basic and applied science. Elsevier, Amsterdam. p. 621–626. doi:10.1016/S0166-2481(08)70378-7

Aylmore, L.A.G., and J.P. Quirk. 1959. Swelling of clay-water systems. Nature 183:1752–1753. doi:10.1038/1831752a0

Babel, U. 1964. Chemische reaktionen an bodendünnschliffen. Leitz-Mitteilungen 3:12–14.

Babel, U. 1972. Fluoreszensmikroscopie in der humusmicroscopie. In: S. Kowalinski, editor, Soil micromorphology. Polska Akademia Nauk, Warsawa, Poland. p. 111–128.

Babel, U. 1975. Micromorphology of soil organic matter. In: J.E.Gieseking, editor, Soil components. Vol. 1, Organic components. Springer, New York. p. 369–473.

Babel, U. 1997. Zur mikromorphologischen untersuchung der organischen substanz des bodens. In: K. Stahr, editor, Mikromorphologische methoden in der bodemkunde. Hohenheimer Bodenkundliche Hefte 40. Universitat Hohenheim, Stuttgart, Germany. p. 7–14.

Babel, U., and H.-J. Vogel. 1989. Zur beurteilung der enchyträen- und collembole-aktivität mit Hilfe von bodendünnschliffen. Pedobiologia 33:167–172.

Baert, G., and E. Van Ranst. 1997. Comparative micromorphological study of representative weathering profiles on different parent materials in the Lower Zaire. In: S. Shoba, M. Gerasimova, and R. Miedema, editors, Soil micromorphology: Studies on soil diversity, diagnostics, dynamics., Wageningen University, Wageningen . p. 28–40.

Bailey, E.H., and R.E. Stevens. 1960. Selective staining of K-Feldspars and Plagioclases on rock slabs and thin sections. Am. Mineral. 45:1020–1025.

Bajwa, I., and D. Jenkins. 1977. The investigation of clay minerals in soil thin-section. In: M. Delgado, editor, Soil micromorphology. Department of Edaphology, Univ. Granada, Granada. p. 3–18.

Bal, L. 1973. Micromorphological analysis of soils. Lower levels in the organization of organic material. Soil Survey Paper 6, Netherlands Soil Survey Institute, Wageningen.

Bal, L. 1977. The formation of carbonate nodules and intercallary crystals in the soil by the earthworm Lumbricus rubellus. Pedobiologia 17:102–106.

Barratt, B.C. 1969. A revised classification and nomenclature of microscopic soil materials with particular reference to organic components. Geoderma 2:257–271. doi:10.1016/0016-7061(69)90026-3

Beckmann, W., and E. Geyger. 1967. Entwurf einer ordnung der natürlichen Hohlraum-, Aggregat- und strukturformen in Boden. In: W.L. Kübiena, editor, Die mikromorphometrische bodenanalyse. Ferd. Enke Verlag, Stuttgart. p. 184–209.

Becze-Deàk, J., R. Langohr, and E.P. Verrecchia. 1997. Small scale secondary CaCO3 accumulations in selected sections of the European loess belt. Morphological forms and potential for paleoenvironmental reconstruction. Geoderma 76:221–252. doi:10.1016/S0016-7061(96)00106-1

Bell, J.D., K. Kulkarni, and M.P. Maraj. 2013. Novel approach to determining unconsolidated reservoir properties: Fabric and flow in oil sands. Unconventional Resources Technology Conference (URTeC), 1–8. Control ID Number: 1581573

Benayas Casares, J. 1963. Partial solution of silica of organic origin in soils. An. Edafol. Agrobiol. 22:623–626.

Bertran, P., J.-P. Texier, and J. Meizeles. 1991. Micromorphology of Atlantic rankers on the coast of northern Portugal. Catena 18:325–343. doi:10.1016/0341-8162(91)90029-W

Betancourt, P.P., and S.E. Peterson. 2009. Thin-section petrography of ceramic materials. Institute for Aegean Prehistory, Archaeological Excavation Manual 2. INSTAP Academic Press, Philadelphia, p. 27.

Bisdom, E.B.A. 1967. Micromorphology of a weathered granite near the Rio de Arose (N.W. Spain). Leidse Geologische Mededelingen 37:33–67.

Bisdom, E.B.A., G. Stoops, J. Delvigne, P. Curmi, and H.-J. Altemüller. 1982. Micromorphology of weathering biotite and its secondary products. Pedologie (Gent) 32:225–252.

Blazejewski, G.A., M.H. Stolt, A.J. Gold, and P.M. Groffman. 2005. Macro- and micromorphology of subsurface carbon in riparian zone soils. Soil Sci. Soc. Am. J. 69:1320–1329. doi:10.2136/sssaj2004.0145

Blokhuis, W.A., S. Slager, and R.G. Van Schagen. 1970. Plasmic fabrics of two Sudan Vertisols. Geoderma 4:127–137. doi:10.1016/0016-7061(70)90040-6

Boggs, S. 2009. Petrology of sedimentary rocks. Second ed. Cambridge Univ. Press, Cambridge, U.K. doi:10.1017/CBO9780511626487

Bordonau, J., and J.J.M. van der Meer. 1994. An example of a kinking microfabric in Upper Pleistocene glaciolacustrine deposits from Llavorsi (Central Southern Pyrenees, Spain). Geol. Mijnb. 73:23–30.

Botanical Society of America. 2008. On-line image collection. http://www.botany.org/plantimages/imagemap.php (Accessed June 2008).

Bouma, J., A. Jongerius, O. Boersma, A. Jager, and D. Schoonderbeek. 1977. The function of different types of macropores during saturated flow through four swelling soil horizons. . Soil Sci. Soc. Am. J. 41:945–950. doi:10.2136/sssaj1977.03615995004100050028x

Bracegirdle, B., and P.H. Miles. 1971. An atlas of plant structure. Heinemann, London.

Brewer, R. 1960. Cutans: Their definition, recognition and classification. J. Soil Sci. 11:280–292. doi:10.1111/j.1365-2389.1960.tb01085.x

Brewer, R. 1964a. Fabric and mineral analysis of soils. John Wiley & Sons, New York.

Brewer, R. 1964b. Classification of plasmic fabric of soil materials. In: A. Jongerius, editor, Soil micromorphology. Elsevier, Amsterdam. p. 95–107.

Brewer, R. 1976. Fabric and mineral analysis of soils. Robert E. Krieger Publ. Co., Huntington, NY.

Brewer, R. 1983. A petrographic approach to soil classification. Sci. Geol. Mem. 73:31–40.

Brewer, R., and S. Pawluk. 1975. Investigations of some soils developed in hummocks of the Canadian SubArctic and Southern Arctic regions: 1 Morphology and micromorphology. Can. J. Soil Sci. 55:301–319. doi:10.4141/cjss75-039

Brewer, R., and J.R. Sleeman. 1960. Soil structure and fabric: Their definition and description. J. Soil Sci. 11:172–185. doi:10.1111/j.1365-2389.1960.tb02213.x

Brewer, R., and J.R. Sleeman. 1963. Pedotubules: Their definition, classification and interpretation. J. Soil Sci. 14:156–166. doi:10.1111/j.1365-2389.1963.tb00941.x

Brewer, R., and J.R. Sleeman. 1964. Glaebules: Their definition, classification and interpretation. J. Soil Sci. 15:66–78. doi:10.1111/j.1365-2389.1964.tb00245.x

Brewer, R., and J.R. Sleeman. 1988. Soil structure and fabric. CSIRO, Australia. doi:10.1071/9780643105492

Brinkman, R., A.G. Jongmans, R. Miedema, and P. Maaskant. 1973. Clay decomposition in seasonally wet, acid soils: Micromorphological, chemical and mineralogical evidences from individual argillans. Geoderma 10:259–270. doi:10.1016/0016-7061(73)90001-3

Brochier, J.E. 1996. Feuilles ou fumiers? Observations sur le rôle des poussières sphérolitiques dans l'interprétation des dépôts archéologiques holocènes. Anthropozoologia 24:19–30.

Brochier, J.E. 2002. Les sédiments anthropiques. Méthodes d'étude et perspectives. In: J.C. Miskovsky, editor, Géologie de la préhistoire: Méthodes, techniques, applications. Géopré Editions, Paris. p. 453–477.

Brochier, J.E., P. Villa, and M. Giacomarra. 1992. Shepherds and sediments: Geo-ethno-archaeology of Pastoral Sites. J. Anthropol. Archaeol. 11:47–102. doi:10.1016/0278-4165(92)90010-9

Brönnimann, D., K. Ismail-Meyer, P. Rentzel, C. Pümpin, and L. Lisá. 2017a. Excrements of herbivores. In: C. Nicosia and G. Stoops, editors, Archaeological soil and sediment micromorphology. John Wiley & Sons, Ltd, Chichester. p. 55–65. doi:10.1002/9781118941065.ch6

Brönnimann, D., C. Pümpin, K. Ismail-Meyer, P. Rentzel, and N. Egüez. 2017b. Excrements of omnivores and carnivores. In: C. Nicosia and G. Stoops, editors, Archaeological soil and sediment micromorphology. John Wiley & Sons, Ltd, Chichester. p. 67–81. doi:10.1002/9781118941065.ch7

Bui, E., and A.R. Mermut. 1989. Quantification of soil calcium carbonates by staining and image analysis. Can. J. Soil Sci. 69:677–682. doi:10.4141/cjss89-066

Bullock, P., and C.P. Murphy. 1976. The microscopic examination of the structure of sub-surface horizons of soils. Outlook Agric. 8:348–354. doi:10.1177/003072707600800607

Bullock, P., P.J. Loveland, and C.P. Murphy. 1975. A technique for selective solution of iron oxides in thin sections of soil. J. Soil Sci. 26:247–249. doi:10.1111/j.1365-2389.1975.tb01948.x

Bullock, P., N. Fedoroff, A. Jongerius, G. Stoops, T. Tursina, and U. Babel. 1985. Handbook for soil thin section description. Waine Research Publications, Wolverhampton, U.K.

Buol, S.W., and S.B. Weed. 1991. Saprolite-soil transformations in the Piedmont and Mountains of North Carolina. Geoderma 51:15–28. doi:10.1016/0016-7061(91)90064-Z

Burns, A., M.D. Pickering, K.A. Green, A.P. Pinder, H. Gestsdóttir, M.R. Usaia, D.R. Brothwell, and B.J. Keely. 2017. Micromorphological and chemical investigation of late-Viking age grave fills at Hofstaðir, Iceland. Geoderma 306:183–194. doi:10.1016/j.geoderma.2017.06.021

Cagauan, B., and G. Uehara. 1965. Soil anisotropy and its relation to aggregate stability. Soil Sci. Soc. Am. Proc. 29:198–200. doi:10.2136/sssaj1965.03615995002900020025x

Canti, M.G. 1997. An investigation of microscopic calcareous spherulites from herbivore dungs. J. Archaeol. Sci. 24:219–231. doi:10.1006/jasc.1996.0105

Canti, M.G. 1998. Origin of calcium carbonate granules found in buried soils and Quaternary deposits. Boreas 27:275–288. doi:10.1111/j.1502-3885.1998.tb01421.x

Canti, M.G. 1999. The production and preservation of faecal spherulites. Animals, Environment and Taphonomy. J. Archaeol. Sci. 26:251–258. doi:10.1006/jasc.1998.0322

Canti, M.G. 2003. Aspects of the chemical and microscopic characteristics of plant ashes found in archaeological soils. Catena 54:339–361. doi:10.1016/S0341-8162(03)00127-9

Canti, M.G. 2017a. Mollusc shell. In: C. Nicosia and G. Stoops, editors, Archaeological soil and sediment micromorphology. John Wiley & Sons, Ltd, Chichester. p. 43–46. doi:10.1002/9781118941065.ch3

Canti, M.G. 2017b. Burnt carbonates. In: C. Nicosia and G. Stoops, editors, Archaeological soil and sediment micromorphology. John Wiley & Sons, Ltd, Chichester. p. 181–188. doi:10.1002/9781118941065.ch22

Canti, M.G. 2017c. Avian eggshell. In: C. Nicosia and G. Stoops, editors, Archaeological soil and sediment micromorphology. John Wiley & Sons, Ltd, Chichester. p. 39–41. doi:10.1002/9781118941065.ch2

Canti, M.G. 2017d. Coal. In: C. Nicosia and G. Stoops, editors, Archaeological soil and sediment micromorphology. John Wiley & Sons, Ltd, Chichester. p. 143–145. doi:10.1002/9781118941065.ch16

Canti, M. G,. 2017e. Charred plant remains. In: Nicosia, C. & Stoops, G. (eds) Archaeological Soil and Sediment Micromorphology. John Wiley & Sons, Ltd, Chichester, pp. 141-142.

Canti, M.G., and J.E. Brochier. 2017a. Faecal spherulites. In: C. Nicosia and G. Stoops, editors, Archaeological soil and sediment micromorphology. John Wiley & Sons, Ltd, Chichester. p. 51–54. doi:10.1002/9781118941065.ch5

Canti, M.G., and J.E. Brochier. 2017b. Plant ash. In: C. Nicosia and G. Stoops, editors, Archaeological soil and sediment micromorphology. John Wiley & Sons, Ltd, Chichester. p. 147–154. doi:10.1002/9781118941065.ch17

Canti, M.G., and T. Piearce. 2003. Morphology and dynamics of calcium carbonate granules produced by different earthworm species. Pedobiologia 47:511–521.

Carpentier, F., and B. Vandermeulen. 2016. High-resolution photography for soil micromorphology slide documentation. Geoarchaeology 31:603–607. doi:10.1002/gea.21563

Catt, J.A. 1989. Relict properties in soils of the central and north-west European temperate region. Catena Suppl. 16:41–58.

Cercone, K.R., and V.A. Pedone. 1987. Fluorescence (photoluminescence) of carbonate rocks: Instrumental and analytical sources of observational error. J. Sediment. Petrol. 57:780–782. doi:10.1306/212F8C42-2B24-11D7-8648000102C1865D

Chadwick, O.A., and W.D. Nettleton. 1994. Quantitative relationships between net volume change and fabric properties during soil evolution. In: A.J. Ringrose-Voase and G.S. Humphreys, editors, Soil micromorphology: Studies in management and genesis, developments in soil science 22. Elsevier Science Publishers, Amsterdam. p. 353–360.

Chiang, S.C., L.T. West, and D.E. Radcliffe. 1994. Morphological properties of surface seals in Georgia soils. Soil Sci. Soc. Am. J. 58:901–910. doi:10.2136/sssaj1994.03615995005800030038x

Choquette, P.W., and L.C. Pray. 1970. Geologic nomenclature and classification of porosity in sedimentary carbonates. Am. Assoc. Pet. Geol. Bull. 54:207–250.

Conway, J., and D. Jenkins. 1977. Application of acetate peels and microchemical staining to soil micromorphology. In: M. Delgado, editor, Soil micromorphology. Department of Edaphology, Univ. Granada, Granada. p. 47–58.

Courty, M.A., P. Goldberg, and R. Macphail. 1989. Soils and micromorphology in archaeology. Cambridge Univ. Press, Cambridge.

Courty, M.A., P. Goldberg, and R. Macphail. 1994. Ancient people- Lifestyles and cultural patterns. Transactions 15th World Congress of Soil Science, Acapulco, Mexico, Vol. 6a: 250–269.

Cremaschi, M., L. Trombino, and A. Zerboni. 2018. Palaeosoils and relict soils: A systematic review. In: G. Stoops, V. Marcelino, and F. Mees, editors, Interpretation of micromorphological features of soils and regoliths. 2nd ed. Elsevier, Amsterdam. p. 863–894. doi:10.1016/B978-0-444-63522-8.00029-2

Curmi, P., A. Soulier, and F. Trolard. 1994. Forms of iron oxides in acid hydromorphic soil environments. Morphology and characterization by selective dissolution. In: A.J. Ringrose-Voase and G.S. Humphreys, editors, Soil micromorphology: Studies in management and genesis. Elsevier, Amsterdam. p. 141–148.

Dalrymple, J.B., and C.Y. Jim. 1984. Experimental study of soil microfabrics induced by isotropic stresses of wetting and drying. Geoderma 34:43–68. doi:10.1016/0016-7061(84)90005-3

Davidson, D.A., S.P. Carter, and T.A. Quine. 1992. An Evaluation of Micromorphology as an Aid to Archaeological Interpretation. Geoarchaeology 7:55–65. doi:10.1002/gea.3340070105

De Coninck, F., D. Righi, J. Maucorps, and A.M. Robin. 1973. Origin and micromorphological nomenclature of organic matter in sandy spodosols. In: G.K. Rutherford, editor, Soil microscopy. The Limestone Press, Kingston (Ont.). p. 263–280.

Deák, J., A. Gebhardt, H. Lewis, M.R. Usai, and H. Lee. 2017. Micromorphology of soils disturbed by vegetation clearance and tillage. In: C. Nicosia and G. Stoops, editors, Archaeological soil and sediment micromorphology. John Wiley & Sons, Ltd, Chichester. p. 231–264. doi:10.1002/9781118941065.ch28

Deer, W.A., R.A. Howie, and J. Zussman. 1971. Rock-forming minerals, Vol. 3, Longmans, London.

Deflandre, G. 1963. Les phytolithaires (Ehrenberg). Protoplasm 57:234–259. doi:10.1007/BF01252058

Delage, A., and H. Lagatu. 1904. Sur la constitution de la terre arable. Comptes Rendus de l'Académie des Sciences 109:1043–1044.

Delvigne, J. 1994. Proposals for classifying and describing secondary microstructures observed within completely weathered minerals. In: A.J. Ringrose-Voase and G.S. Humphreys, editors, Soil micromorphology: Studies in management and genesis. Elsevier, Amsterdam. p. 333–342.

Delvigne, J.E. 1998. Atlas of micromorphology of mineral alteration and weathering. The Canadian Mineralogist, Special Publ. 3.

Delvigne, J., E.B.A. Bisdom, J. Sleeman, and G. Stoops. 1979. Olivines, their pseudomorphs and secondary products. Pedologie (Gent) 29: 247–309.

De Paepe, P., and G. Stoops. 2007. A classification of tephra in volcanic soils. A tool for soil scientists. In: O. Arnalds, F. Bartoli, P. Buurman, H. Oskarsson, G. Stoops, and E. Garcia-Rodeja, editors, Soils of volcanic regions in Europe. Springer, Berlin, Heidelberg, New York. p. 119–125. doi:10.1007/978-3-540-48711-1_12

Devos, Y., L. Vrydaghs, A. Degraeve, and K. Fechner. 2009. Palaeoenvironmental research on the site of rue the Dinant (Brussels): An interdisciplinary study of dark earth. Catena 78:270–284. doi:10.1016/j.catena.2009.02.013

Dickson, J.A.D. 1965. A modified staining technique for carbonates in thin sections. Nature 205:587. doi:10.1038/205587a0

Dobrovol'ski, G.V., editor. 1983. A methodological manual of soil micromorphology. ITC-Gent, Ghent, Belgium.

Donaldson, C., and C.M.B. Henderson. 1988. A new interpretation of round embayements in quartz crystals. Mineral. Mag. 52:27–33. doi:10.1180/minmag.1988.052.364.02

Dorronsoro, C., E. Ortega, and M. Delgado. 1978a. The use of fluorescent paints in micromorphometric studies by automatic image analysis systems. In: M. Delgado, editor, Soil micromorphology. University of Granada, Spain. p. 1253–1268.

Dorronsoro, C., E. Ortega, and M. Delgado. 1978b. Some facts concerning the accuracy of soil measuring porosity by image analysis systems. In: M. Delgado, editor, Soil micromorphology. University of Granada, Spain. p. 1269–1302.

Douglas, L.A., editor. 1990. Soil micromorphology: A basic and applied science. Developments in Soil Science 19. Elsevier, Amsterdam.

Douglas, L.A., and M.L. Thompson, editors. 1985. Soil micromorphology and soil classification. SSSA Spec. Publ. 15, Madison, WI.

Dravis, J.J., and D.A. Yurewicz. 1985. Enhanced carbonate petrography using fluorescence microscopy. J. Sediment. Petrol. 55:795–804.

Drees, L.R., and M.D. Ransom. 1994. Light microscopic techniques in quantitative soil mineralogy. In: J.E. Amonette and L.W. Zelazny, editors, Quantitative methods in soil mineralogy. SSSA, Madison, WI. p. 137–176.

Drees, L.R., L.P. Wilding, N.E. Smeck, and A.L. Senkayi. 1989. Silica in soils: Quartz and disordered silica polymorphs. In: J.B. Dixon and S.B. Weed, editors, Minerals in soil environments. SSSA, Madison, WI. p. 913–974.

Driessen, P.M., and R. Schoorl. 1973. Mineralogy and morphology of salt efflorescences on saline soils in the Great Konya Basin, Turkey. J. Soil Sci. 24:436–442. doi:10.1111/j.1365-2389.1973.tb02310.x

Ducloux, J., P. Butel, and T. Dupuis. 1984. Microséquence minéralogique des carbonates de calcium dans une accumulation carbonatée sous galets calcaires, dans l'ouest de la France. Pedologie (Gent) 34:161–177.

Dumanski, J., and R.J. St. Arnaud. 1966. A micropedological study of eluvial horizons. Can. J. Soil Sci. 46:287–292. doi:10.4141/cjss66-044

Dunham, R.J. 1962. Classification of carbonate rocks according to depositional texture. In: W.E. Ham, editor, Classification of carbonate rocks. Vol. 1. American Association of Petroleum Geologists, Tulsa, OK. p. 108–121.

Ekdale, A.A., and R.G. Bromley. 1991. Analysis of composite ichnofabrics: An example in Uppermost Cretaceous chalk in Denmark. Palaios 6:232–249. doi:10.2307/3514904

Embrechts, J., and G. Stoops. 1986. Relations between microscopical features and analytical characteristics of a soil catena in a humid tropical climate. Pedologie (Gent) 36:315–328.

Esau, K. 1977. Anatomy of seed plants, 2nd ed. John Wiley & Sons, New York.

Eswaran, H., and C. Baños. 1976. Related distribution patterns in soils and their significance. An. Edafol. Agrobiol. 35:33–45.

Eswaran, H., C. Sys, and E.C. Sousa. 1975. Plasma infusion. A pedological process of significance in the humid tropics. An. Edafol. Agrobiol. 34:665–674.

Evamy, B.D. 1963. The application of a chemical staining technique to a study of dedolomitisation. Sedimentology 2:164–170. doi:10.1111/j.1365-3091.1963.tb01210.x

Fedoroff, N. 1997. Clay illuviation in red Mediterranean soils. Catena 28:171–189. doi:10.1016/S0341-8162(96)00036-7

Fedoroff, N., and M.A. Courty. 1994. Organisation du sol aux échelles microscopiques. In: M. Bonneau and B. Souchier, editors, Pédologie. 2. Constituants et propriétés du sol. Masson, Paris. p. 349–375.

Fedoroff, N., M.A. Courty, and Z. Guo. 2018. Palaeosoils and relict soils: A conceptual approach. In: G. Stoops, V. Marcelino, and F. Mees, editors, Interpretation of micromorphological features of soils and regoliths. 2nd ed. Elsevier, Amsterdam. p. 821–862. doi:10.1016/B978-0-444-63522-8.00028-0

Fettes, D., and J. Desmons. 2007. Metamorphic rocks. A classification and glossary of terms. Recommendations of International Union of Geological Sciences Subcommission on the Systematics of Metamorphic Rocks. Cambridge Univ. Press, Cambridge.

FitzPatrick, E.A. 1956. An indurated soil horizon formed by permafrost. J. Soil Sci. 7:248–254. doi:10.1111/j.1365-2389.1956.tb00882.x

FitzPatrick, E.A. 1984. Micromorphology of soils. Chapman and Hall, London. doi:10.1007/978-94-009-5544-8

FitzPatrick, E.A. 1989. The use of the term Birefringence in soil micromorphology. Soil Sci. 147:357–360. doi:10.1097/00010694-198905000-00006

FitzPatrick, E.A. 1993. Soil microscopy and micromorphology. John Wiley & Sons, Chichester, UK.

Flügel, E. 2010. Microfacies of carbonate rocks. Analysis, interpretation and application. Springer, Heidelberg.

Folk, R.L. 1962. Spectral subdivision of limestone types. In: W.E. Ham, editor, Classification of carbonate rocks. Vol. 1. American Association of Petroleum Geologists, Tulsa, OK,. p. 62–84.

Fox, C.A., and L.E. Parent. 1993. Micromorphological methodology for organic soils. In: M.R. Carver, editor, Soil sampling and methods of analysis. Lewis Publishers, Boca Raton, FL. p. 473–485.

Fox, C.A., R.K. Guertin, E. Dickson, S. Sweeney, R. Protz, and A.R. Mermut. 1993. Micromorphological methodology for inorganic soils. In: M.R. Carver, editor, Soil sampling and methods of analysis. Lewis Publishers, Boca Raton. p. 683–709.

Friedman, G.M. 1959. Identification of carbonate staining methods. J. Sediment. Petrol. 29:87–97.

Friedman, G.M. 1965. Terminology of crystallization textures and fabrics in sedimentary rocks. J. Sediment. Petrol. 35:643–655.

Galopin, R., and N.F.M. Henry. 1972. Microscopic study of opaque minerals. W. Heffer and Sons, Ltd. Cambridge.

Gay, P. 1982. An introduction to crystal optics. Longman, London.

Gebhardt, A., and R. Langhor. 1999. Micromorphological study of construction materials and living floors in the medieval motte of Werken (West Flanders, Belgium). Geoarchaeology 14:595–620. doi:10.1002/(SICI)1520-6548(199910)14:7<595::AID-GEA1>3.0.CO;2-Q

Gerasimova, M.I. 1994. Integral micromorphological characteristics of soils (Morphotypes of Horizons and Morphons). Eurasian Soil Sci. 26:65–76.

Gerasimova, M., and M. Lebedeva-Verba. 2018. Topsoils. In: G. Stoops, V. Marcelino, and F. Mees, editors, Interpretation of micromorphological features of soils and regoliths. 2nd ed. Elsevier, Amsterdam. p. 413–538.

Gerasimova, M.I., S.V. Gubin, and S.A. Shoba. 1996. Soils of Russia and adjacent countries: Geography and micromorphology. Wageningen University, Wageningen, The Netherlands.

Gifford, E.M., and A.S. Foster. 1989. Morphology and evolution of vascular plants. Third ed. Freeman, New York.

Goldberg, P., and V. Aldeias. 2018. Why does (archaeological) micromorphology have such little traction in (geo)archaeology? Archaeological and Anthropological Sciences 10: 269–278.

Goldberg, P., and R.I. Macphail. 2006. Practical and theoretical geoarchaeology. Blackwell Publishing, Oxford.

Golyeva, A.A. 2008. Microbiomorfic analysis as tool for natural and anthropogenic landscape investigation. Russian Academy of Science, Moscow.

Gutiérrez-Castorena, M.C., E.V. Gutierrez-Castorena, T. Gonzalez-Vargas, A. Ortiz-Solorio, E. Suástegui-Méndez, L. Cajuste-Bontemps and M.N. Rodríguez-Mendoza. 2018. Thematic micro-maps of soil components using high-resolution spatially referenced mosaics from whole soil thin sections and image analysis. Eur. J. Soil Sci. 69:217–231. doi:10.1111/ejss.12506

Gutiérrez-Rodriguez, M., M. Toscano, and P. Goldberg. 2018. High-resolution dynamic illustrations in soil micromorphology: A proposal for presenting and sharing primary research data in publication. Journal of Archaeological Science-Reports 20:565–575. doi:10.1016/j.jasrep.2018.05.025

Harrell, J.A. 1981. Measurement errors in the thin section analysis of grain packing. J. Sediment. Petrol. 51:674–676. doi:10.1306/212F7D60-2B24-11D7-8648000102C1865D

Harrell, J.A. 1984. A visual comparator for degree of sorting in thin and plane sections. J. Sediment. Petrol. 54:646–650. doi:10.2110/jsr.54.646

Harrell, J.A., and K.A. Eriksson. 1979. Empirical conversion equations for thin-section and sieve derived distribution parameters. J. Sediment. Petrol. 49:273–280.

Hartmann, C., D. Tessier, and L.P. Wilding. 1992. Simultaneous use of transmitted and incident ultraviolet light in describing soil microfabrics. Soil Sci. Soc. Am. J. 56:1867–1970. doi:10.2136/sssaj1992.03615995005600060036x

Herrero, J., and J. Porta. 1987. Gypsiferous soils in the North-East of Spain. In: N. Fedoroff, L.M. Bresson, and M.A. Courty, editors, Micromorphologie des sols- Soil micromorphology. AFES, Paris. p. 186–192.

Hiemstra, J.F., and J.J.P. van der Meer. 1997. Pore controlled grain fracturing as an indicator for subglacial shearing in tills. J. Glaciol. 43(145):446–454. doi:10.1017/S0022143000035036

Hill, I.D. 1970. The use of orientation diagrams in describing plasmic fabrics in soil materials. J. Soil Sci. 21:184–187. doi:10.1111/j.1365-2389.1970. tb01166.x

Hill, I.D. 1981. A method for the quantitative measurement of anisotropic plasma in soil thin sections. J. Soil Sci. 32:461–464. doi:10.1111/j.1365-2389.1981. tb01722.x

Hodgson, J.M., editor. 1976. Soil survey field handbook. Soil Survey Tech. Monogr. 5, Soil Survey, Harpenden, England.

Holmes, A. 1927. Petrographic methods and calculations. Murby, London.

Houghton, H.F. 1980. Refined technique for staining plagioclase and alkali feldspars in thin section. J. Sediment. Petrol. 50:629–631. doi:10.1306/212F7A7C-2B24-11D7-8648000102C1865D

Huijzer, A.S. 1993. Cryogenic microfabrics and macrostructures: Interrelation, processes, and paleoenvironmental significance. Ph.D. diss. Vrije Universiteit Amsterdam.

Hutchison, Ch.S. 1974. Laboratory handbook of petrographic techniques. J. Wiley & Sons, New York.

International Committee for Phytolith Taxonomy. 2019. International code for phytolith nomenclature 2.0. Annals of Botany 124(2):189–199.

Ismail-Meyer, K. 2017. Plant remains. In: C. Nicosia and G. Stoops, editors, Archaeological soil and sediment micromorphology. John Wiley & Sons Ltd, Chichester. p. 121–135. doi:10.1002/9781118941065.ch13

Ismail-Meyer, K., M. Stolt, and D. Lindbo. 2018. Soil organic matter. In: G. Stoops, V. Marcelino, and F. Mees, editors, Interpretation of micromorphological features of soils and regoliths. 2nd ed. Elsevier, Amsterdam. p. 471–512.

Jaillard, B., A. Guyon, and A.F. Maurin. 1991. Structure et composition of calcified roots, and their identification in calcareous soils. Geoderma 50:197–210. doi:10.1016/0016-7061(91)90034-Q

Jambor, J.L., and D.J. Vaughan. 1990. Advanced microscopic studies of ore minerals. Mineralogical association of Canada Short Course, Ottawa, ON.

Jeanroy, E., J.L. Rajot, P. Pillon, and A.J. Herbillon. 1991. Differential dissolution of hematite and goethite in dithionite and its application on soil yellowing. Geoderma 50:79–94. doi:10.1016/0016-7061(91)90027-Q

Jenkins, D.A. 1994. Interpretation of interglacial cave sediments from hominid site in North Wales: Translocation of Ca-Fe-phosphates. In: A.J. Ringrose-Voase and G.S. Humphreys, editors, Soil micromorphology: Studies in management and genesis. Elsevier, Amsterdam. p. 293–305.

Jenkins, E. 2009. Phytolith taphonomy: A comparison of dry ashing and acide extraction on the breakdown of conjoined phytolith formed in Triticum durum. J. Archaeol. Sci. 36:2402–2407. doi:10.1016/j.jas.2009.06.028

Jim, C.Y. 1988a. A classification of soil microfabrics. Geoderma 41:315–325. doi:10.1016/0016-7061(88)90067-5

Jim, C.Y. 1988b. Microscopic-photometric measurement of polymodal clay orientation using circular-polarized light and interference colours. Appl. Clay Sci. 3:307–321. doi:10.1016/0169-1317(88)90022-1

Jim, C.Y. 1990. Stress, shear deformation and micromorphological clay orientation: A synthesis of various concepts. Catena 17:431–447. doi:10.1016/0341-8162(90)90044-E

Johnson, M.A. 1994. Thin section grain size analysis revisited. Sedimentology 41:985–999. doi:10.1111/j.1365-3091.1994.tb01436.x

Jongerius, A. 1957. Morfologische onderzoekingen over de Bodemstructuur. Bodemkundige Studies No. 2. Mededelingen van de Stichting voor Bodemkartering. Wageningen, The Netherlands.

Jongerius, A., and G.K. Rutherford, editors. 1979. Glossary of soil micromorphology. Pudoc, Wageningen.

Jongmans, A.G., N. Van Breemen, U. Lundström, P.A.W. van Hees, R.D. Finlay, M. Srinivasan, T. Uestam, R. Giesler, P.-A. Melkerud, and M. Olsson. 1997. Rock-eating fungi. Nature 389:682–683. doi:10.1038/39493

Kaczorek, D., L. Vrydaghs, Y. Devos, A. Petó, and W. Effland. 2018. Biogenic siliceous features. In: G. Stoops, V. Marcelino, and F. Mees, editors, Interpretation of micromorphological features of soils and regoliths. 2nd ed. Elsevier, Amsterdam. p. 157–176. doi:10.1016/B978-0-444-63522-8.00007-3

Kapur, S., C. Karaman, E. Akca, M. Aydin, U. Dinc, E.A. FitzPatric, M. Pagliai, D. Kalmar, and A.R. Mermut. 1997. Similarities and differences of the spheroidal microstructure in Vertisols from Turkey and Israel. Catena 28:297–311. doi:10.1016/S0341-8162(96)00044-6

Karale, R.L., E.B.A. Bisdom, and A. Jongerius. 1974. Micromorphological studies on diagnostic subsurface horizons of some of the alluvial soils in the Meerut district of Uttar Pradesh. J. Indian Soc. Soil Sci. 22:70–76.

Karkanas, P. 2017. Guano. In: C. Nicosia and G. Stoops, editors, Archaeological soil and sediment micromorphology. John Wiley & Sons, Ltd, Chichester. p. 83–89. doi:10.1002/9781118941065.ch8

Karkanas, P., and P. Goldberg. 2018. Phosphatic features. In: G. Stoops, V. Marcelino, and F. Mees, editors, Interpretation of micromorphological features of soils and regoliths. 2nd ed. Elsevier, Amsterdam. p. 323–346. doi:10.1016/B978-0-444-63522-8.00012-7

Kemp, R.A. 1999. Soil micromorphology as a technique for reconstructing paleoenvironmental change. In: A.S. Singh Vi and E. Derbyshire, editors, Paleoenvironmental reconstruction in arid lands. Balkema Pub., Netherlands. p. 41–71.

Khormali, F., A. Abtahi, and G. Stoops. 2006. Micromorphology of calcitic features. Geoderma 132:31–46. doi:10.1016/j.geoderma.2005.04.024

Kooistra, M.J. 2015. Descripción de los componentes orgánicos del suelo. In: J.C. Loaiza, G. Stoops, R. Poch, and M. Casamitjana, editors, Manual de micromorfología de suelos y técnicas complementarias. Fondo Editorial Pascual Bravo, Medellin, Colombia. p. 261–292.

Koopman, G.J. 1988. "Waterhard": A hard brown layer in sand below peat, The Netherlands. Geoderma 42:147–157. doi:10.1016/0016-7061(88)90030-4

Korina, N.A., and M.A. Faustova. 1964. Microfabric of modern and old moraines. In: A. Jongerius, editor, Soil micromorphology. Elsevier, Amsterdam. p. 333–338.

Krebs, M., A. Kretzschmar, U. Babel, J. Chadoeuf, and M. Goulard. 1994. Investigations on distribution patterns in soils: Basic and relative distributions of roots, channels, and cracks. In: A.J. Ringrose-Voase and G.S. Humphreys, editors, Soil micromorphology: Studies in management and genesis. Elsevier, Amsterdam. p. 437–449.

Krumbein, W.C., and L.L. Sloss. 1963. Stratigraphy and sedimentation. W.H. Freeman & Co., San Francisco, CA.

Kubiëna, W.L. 1931. Micropedologische studien. Archief für Pflanzenbau 5:615–648.

Kubiëna, W.L. 1938. Micropedology. Collegiate Press, Ames, IA.

Kubiëna, W.L. 1948. Entwicklungslehre des Bodens. Springer-Verlag, Wien. doi:10.1007/978-3-7091-7714-3

Kubiëna, W.L. 1953. The soils of Europe. Thomas Murby & Co., London.

Kubiëna, W.L. 1956. Zur Mikromorfologie, Systematik und Entwicklung der rezenten und fossilen Lössboden. Eiszeitalter Ggw. 7:102–112.

Kubiëna, W.L. 1970. Micromorphological features of soil geography. Rutgers Univ. Press, New Brunswick, NJ.

Kühn, P., J. Aguilar, R. Miedema, and M. Bronnikova. 2018. Textural features and related horizons. In: G. Stoops, V. Marcelino, and F. Mees, editors, Interpretation of micromorphological features of soils and regoliths. 2nd ed. Elsevier, Amsterdam. p. 377–423. doi:10.1016/B978-0-444-63522-8.00014-0

Lafeber, D. 1964. Soil fabric and soil mechanics. In: A. Jongerius, editor, Soil micromorphology. Elsevier, Amsterdam. p. 351–360.

Le Maitre, R.W., editor. 2002. Igneous rocks. A classification and glossary of terms. Recommendations of International Union of Geological Sciences Subcommission on the Systematics of Igneous Rocks. Second ed. Cambridge Univ. Press, Cambridge. doi:10.1017/CBO9780511535581

Ligouis, B. 2017. Reflected light. In: C. Nicosia and G. Stoops, editors, Archaeological soil and sediment micromorphology. John Wiley & Sons, Ltd, Chichester. p. 461–470. doi:10.1002/9781118941065.ch44

Loaiza, J.C., G. Stoops, R. Poch, and M. Casamitjana, editors. 2015. Manual de micromorfología de suelos y técnicas complementarias. Fonda Editorial Pascual Bravo, Medelin, Colombia.

Longiaru, S. 1987. Visual comparators for estimating the degree of sorting from plane and thin sections. J. Sediment. Petrol. 57:791–794. doi:10.1306/212F8C60-2B24-11D7-8648000102C1865D

MacKenzie, W.S., and A.E. Adams. 1993. A colour atlas of rocks and minerals in thin section. Manson Publ. Ltd., London.

MacKenzie, W.C., and C. Guilford. 1980. Atlas of rocks-forming minerals in thin sections. Longman, Essex, U.K.

MacKenzie, W.S., C.H. Donaldson, and C. Guilford. 1982. Atlas of igneous rocks and their textures. Longman, Essex.

Macphail, R.I. 2008. Soils and archaeology. In: D.M. Pearsall, editor, Encyclopedia of archaeology. Academic Press, New York. p. 2064–2072. doi:10.1016/B978-012373962-9.00290-9

Macphail, R.I. 2014. Archaeological soil micromorphology. In: C. Smith, editor, Encyclopedia of global archaeology. Springer, Amsterdam. p. 356–364. doi:10.1007/978-1-4419-0465-2_227

Macphail, R.I., and P. Goldberg. 2018. Archaeological materials. In: G. Stoops, V. Marcelino, and F. Mees, editors, Interpretation of micromorphological features of soils and regoliths. 2nd ed. Elsevier, Amsterdam. p. 779–819. doi:10.1016/B978-0-444-63522-8.00027-9

Macphail, R.I., M.A. Courty and P. Goldberg. 1990. Soil micromorphology in archaeology. Endeavour, New Series 14, 163–171. doi:10.1016/0160-9327(90)90039-T

Madella, M., and C. Lancelotti. 2012. Taphonomy and phytoliths: A user manual. Quat. Int. 275:76–83. doi:10.1016/j.quaint.2011.09.008

Magaldi, D. 1974. Caratteri a modalita dell orientamento delle argile nel'orizzonte B di alcuni suoli. Atti. Soc. Tosc. Scl. Net. Mem. Series A 81:152–166.

Maier-Kühne, H.-M., and U. Babel. 1984. Berücksichtigung des Holmes-Effekts bei stereologischen Messungen an Dünnschliffen. Mitt. Dtsch. Bodenkundl. Gesellsch. 39:79–84.

Marcelino, V., V. Cnudde, S. Van Steeland, and F. Caro. 2007. An evaluation of 2D-image analysis techniques for measuring soil microporosity. Eur. J. Soil Sci. 58:133–140. doi:10.1111/j.1365-2389.2006.00819.x

Marchall, D. 1988. Cathodoluminescence of geological materials. Unwin Hyman Ltd., London.

Maritan, L. 2017. Ceramic materials. In: C. Nicosia and G. Stoops, editors, Archaeological soil and sediment micromorphology. Wiley, Chichester. p. 205–212. doi:10.1002/9781118941065.ch25

McKeague, J.A., R.K. Guertin, K.W. Valentine, J. Belisle, G.A. Bourbeau, W. Michalyna, L. Hopkins, L. Howell, F. Page, and L.M. Bresson. 1980. Variability of estimates of illuvial clay in soils by micromorphology. Soil Sci. 129:386–388. doi:10.1097/00010694-198006000-00009

McLoughlin, N., H. Furnes, N.R. Banerjee, H. Staudigel, K. Muehlenbachs, M. de Wit, and M.J. Van Kranendonk. 2008. Micro-bioerosion in volcanic glass: Extending the ichnofossil record to Archaean basaltic crust. In: M. Wisshak and L. Tapanila, editors, Current developments in bioerosion. Springer, Berlin. p. 371–396. doi:10.1007/978-3-540-77598-0_19

McLoughlin, N., H. Staudigel, H. Furnes, B. Eickmann, and M. Ivarsson. 2010. Mechanisms of microtunneling in rock substrates: Distinguishing endolithic biosignatures from abiotic microtunnels. Geobiology 8:245–255. doi:10.1111/j.1472-4669.2010.00243.x

Mees, F., and G. Stoops. 1991. Mineralogical study of salt efflorescences on soils of the Jequetepeque Valley, northern Peru. Geoderma 49:255–272. doi:10.1016/0016-7061(91)90079-9

Mees, F., and T.V. Tursina. 2018. Salt minerals in saline soils and salt crusts. In: G. Stoops, V. Marcelino, and F. Mees, editors, Interpretation of micromorphological features of soils and regoliths. 2nd ed. Elsevier, Amsterdam. p. 289–321. doi:10.1016/B978-0-444-63522-8.00011-5

Melgarejo, J.-C., editor. 1997. Atlas de associaciones minerales en lamina delgada. Ed. Univ., Barcelona, Spain.

Menzies, J., and J.J.M. van der Meer. 2018. Micromorphology and microsedimentology of glacial sediments. In: J. Menzies and J. van der Meer, editors, Past glacial environments. 2nd ed. Elsevier, Amsterdam. p. 753–806. doi:10.1016/B978-0-08-100524-8.00036-1

Mermut, A., and R.J. St.Arnaud. 1981. Microband fabric in seasonally frozen soils. Soil Sci. Soc. Am. J. 45:578–586. doi:10.2136/sssaj1981.03615995004500030029x

Miedema, R., and A.R. Mermut. 1990. Soil micromorphology: An annotated bibliography 1968- 1986. CAB, Wallingford, U.K.

Minashina, N.G. 1958. Optically oriented clays in soils. Sov. Soil Sci. 4:424–430.

Mitchell, J.K. 1956. The fabric of natural clays and its relation to engineering properties. Highw. Res. 35:693–713.

Moon, C.F. 1972. The microstructure of clay sediments. Earth Sci. Rev. 8:303–321. doi:10.1016/0012-8252(72)90112-2

Moran, C.J., A.J. Koppi, B.W. Murphy, and A.B. McBratney. 1988. Comparison of the macropore structure of a sandy loam surface horizon subjected to two tillage treatments. Soil Use Manage. 4:96–102. doi:10.1111/j.1475-2743.1988.tb00743.x

Morgenstern, N.R., and J.S. Tchalenko. 1967a. The optical determination of preferred orientation in clays and its application to the study of microstructure in consolidated kaolin I + II. Proc. R. Soc. Lond. A 300:218–234. doi:10.1098/rspa.1967.0167

Morgenstern, N.R., and J.S. Tchalenko. 1967b. Microscopic structures in kaolin subjected to direct shear. Geotechnique 17:309–328. doi:10.1680/geot.1967.17.4.309

Morrás, H. 1973. L'identification des matériaux carbonatés par coloration différentielle. Technical note 1. Bodemmineralogie en mikromorfologie, Rijksuniversiteit, Gent.

Morrás, H. 2015. Porosidad y microestructura de suelos. In: J.C. Loaiza, G. Stoops, R. Poch, and M. Casamitjana, editors, Manual de micromorphología de suelos y técnicas complementarias. Fonda Editorial Pascual Bravo, Medelin, Colombia. p. 205–260.

Murphy, C.P. 1986. Thin section preparation of soils and sediments. A B Academic Publ., Berkhamsted.

Murphy, C.P., and R.A. Kemp. 1984. The over-estimation of clay and the underestimation of pores in soil thin sections. J. Soil Sci. 35:481–495. doi:10.1111/j.1365-2389.1984.tb00305.x

Murphy, C.P., P. Bullock, and R.H. Turner. 1977. The measurement and characterisation of voids in thin sections by image analysis. Part 1. Principles and techniques. J. Soil Sci. 28:498–508. doi:10.1111/j.1365-2389.1977.tb02258.x

Nahon, D.B. 1991. Introduction to the petrology of soils and chemical weathering. J. Wiley & Sons, New York.

Nesse, W.D. 2003. Introduction to optical mineralogy. Third ed. Oxford Univ. Press, Oxford.

Nicosia, C., and G. Stoops, editors. 2017. Archaeological soil and sediment micromorphology. John Wiley & Sons, Ltd, Chichester. doi:10.1002/9781118941065

Pagel, M., V. Barbin, P. Blanc, and D. Ohnenstetter. 2000. Cathodoluminescence in Geosciences. Springer Verlag.

Pagliai, M., and P. Sequi. 1982. A comparision between two treatments for removal of iron oxides from thin sections of a soil. Can. J. Soil Sci. 62:533–535. doi:10.4141/cjss82-058

Pape, Th. 1974. The application of circular polarized light in soil micromorphology. Neth. J. Agric. Sci. 22:31–36.

Parfenoff, A., C. Pomerol, and J. Tourenq. 1970. Les Minéraux en Grains. Méthodes d'Etude et de Détermination. Masson, Paris

Parfenova, E., and E.A. Yarilova. 1965. Mineralogical investigations in soil science. (In Russian, translated into English). Israel Programme for Scientific Translations, Jerusalem.

Peckett, A. 1992. The colours of opaque minerals. J. Wiley & Sons. New York.

Perkins, D., and K.R. Henke. 2000. Minerals in thin sections. Prentice Hall, NJ.

Pettijohn, F.J. 1957. Sedimentary rocks. Harper and Row, New York.

Pettijohn, F.J., P.E. Potter, and R. Siever. 2012. Sand and sandstone. Second ed. Springer Verlag, Berlin.

Phillips, W.R. 1971. Mineral optics: Principles and techniques. W.H. Freedman Co., San Francisco, CA.

Phillips, W.R., and D.T. Griffen. 1981. Optical mineralogy: The nonopaque minerals. W.H. Freedman Co., San Francisco.

Pichler, H., and C. Schmitt-Riegraf. 1997. Rock-forming minerals in thin sections. Chapman and Hall, London. doi:10.1007/978-94-009-1443-8

Piperno, D.R. 1988. Phytolith analysis-An archaeological and geological perspective. Academic Press, London.

Postma, J., and H.-J. Altemüller. 1990. Bacteria in thin soil sections stained with fluorescent brightener calcofluor white M2R. Soil Biol. Biochem. 22:89–96. doi:10.1016/0038-0717(90)90065-8

Powers, M.C. 1953. A new roundness scale for sedimentary particles. J. Sediment. Petrol. 23:117–119. doi:10.1306/D4269567-2B26-11D7-8648000102C1865D

Puhan, D. 1994. Anleitung zur Dünnschliffmikroskopie. Enke Verlag, Stuttgart.

Quinn, P.S. 2013. Ceramic petrography: The interpretation of archaeological pottery and related artefacts in thin sections. Archaeopress, Oxford.

Raith, M.M., P. Raase, and J. Reinhardt. 2011. Thin section microscopy. Mineralogical Society of America, Chantilly, VA. http://www.minsocam.org/msa/openaccess_publications/#Guide (Accessed 11 Sept. 2019).

Rassineux, F., D. Beaufort, A. Meunier, and A. Bouchet. 1987. A method of coloration by fluorescein aqueous solution for thin-section microscopic observation. J. Sediment. Petrol. 57:782–783. doi:10.1306/212F8C47-2B24-11D7-8648000102C1865D

Ringrose-Voase, A.J. 1991. Micromorphology of soil structure: Description, quantification, application. Aust. J. Soil Res. 29:777–813. doi:10.1071/SR9910777

Romans, J.C., J. Stevens, and L. Robertson. 1966. Alpine soils of northeast Scotland. J. Soil Sci. 17:184–199. doi:10.1111/j.1365-2389.1966.tb01465.x

Rovner, I. 1971. Potential of opal phytoliths for use in paleoecological reconstruction. Quat. Res. 1:343–359. doi:10.1016/0033-5894(71)90070-6

Ruark, G.A., P.L.M. Veneman, D.L. Mader, and P.F. Waldron. 1982. Use of circular polarization on soil thin sections to distinguish voids from mineral grains. Soil Sci. Soc. Am. J. 46:880–882. doi:10.2136/sssaj1982.03615995004600040043x

Runge, F. 1998. The effect of dry oxidation temperatures (500 °C–800 °C) and of natural corrosion on opal phytoliths. In: Abstracts of the Deuxième Congrès International de Recherches sur les Phytolithes, Aix-en-Provence. p. 73.

Runge, F. 1999. The opal phytolith inventory of soils in central Africa-Quantities, shapes, classification and spectra. Rev. Palaeobot. Palynol. 107:23–53. doi:10.1016/S0034-6667(99)00018-4

Russell, R.D., and R.E. Taylor. 1937. Roundness and shape of Mississippi River sands. J. Geol. 45:225–267. doi:10.1086/624526

Sander, B. 1948. Einfuhrung in die Gefügekunde der geologischen Körper. Teil I. Algemeine Gefügekunde und Arbeiten im Bereich Handstuck bis Profil. Springer-Verlag, Wien.

Sander, B. 1950. Einfuhrung in die Gefügekunde der geologischen Körper. Teil II. Die Korngefüge. Springer Verlag, Wien. doi:10.1007/978-3-7091-7759-4

Sander, B. 1970. An introduction to the study of fabrics of geological bodies. Pergamon Press, Oxford.

Schnütgen, A., and H. Späth. 1983. Mikromorphologische Sprengung von Quarzkörnern durch Eisenverbindungen in tropischen Böden. Zeitschrift für Geomorphologie 48:17–34.

Schoch, W.H., I. Heller, F.H. Schweingruber, and F. Kienast. 2004. Wood anatomy of central European species. Eidg. Forschungsanstalt für Wald, Schnee und Landschaft, Zurich, Switzerland. http://www.woodanatomy.ch (Accessed 11 Sept. 2019).

Schoch, W.H., B. Pawlik, and F.H. Schweingruber. 1988. Botanical macro-remains: An atlas for the determination of frequently encountered and ecologically important plant remains. Paul Haupt Publishers, Bern.

Schoenberger, P.J., D.A. Wysocki, E.C. Benham, and W.D. Broderson. 1998. Field book for describing and sampling soils. Natural Resources Conservation Service, National Soil Survey Center, Lincoln, NE.

Scholle, P.A., and D.S. Ulmer-Scholle. 2003. A color guide to the petrography of carbonate rocks: Grains, textures, porosity, diagenesis. American Association of Petroleum Geologists Tulsa, OK.

Schumacher, R., and H.-U. Schmincke. 1991. Internal structure and occurrence of accretionary lapilli- a case study at Laacher See Volcano. Bull. Volcanol. 53:612–634. doi:10.1007/BF00493689

Schumacher, R., and H.-U. Schmincke. 1995. Models for the origin of accretionary lapilli. Bull. Volcanol. 56:626–639. doi:10.1007/BF00301467

Schweingruber, F.H. 1982. Microscopic wood anatomy, structural variability of stems and twigs in recent and sufossil woods from Central Europe. Fluck-Wirth, Teufen, Switzerland.

Sen, R., and A.D. Mukherjee. 1972. The relation of petrofabrics with directional orientation of mineral grains from soil parent materials. An example from Norway. Soil Sci. 113:57–58. doi:10.1097/00010694-197201000-00011

Shahack-Gross, R. 2015. Archaeological micromorphology self-evaluation exercise. Geoarchaeology 31:49–57. doi:10.1002/gea.21536

Simões de Castro, S. and M. Cooper. 2019. Fundamentos de micromorfologia de solos. Viçosa, Brazil: Sociedade Brasileira de Ciéncia do Solo.

Simkiss, K., and K.M. Wilbur. 1989. Biomineralization cell biology and mineral deposition. Academic Press, San Diego, CA.

Smaill, S.J., P.W. Clinton, R.B. Allen, and M.R. Davis. 2014. New evidence indicates the coarse soil fraction is of greater relevance to plant nutrition than previously suggested. Plant Soil 374:371–379. doi:10.1007/s11104-013-1898-3

Smart, P. 1969. Soil structure in the electron microscope. In: Structure, solid mechanics and engineering design. Proc. Southampton Civil Engin. Mat. Conf. Part 1. p. 249–255.

Soil Survey Staff. 1975. Soil taxonomy: A basic system of soil classification for making and interpreting soil surveys. USDA Agric. Handbook N° 436, Washington, D.C.

Soil Survey Staff. 1993. Soil survey manual. Handbook 18. USDA, Washington, D.C.

Stephan, S. 1972. In: Mikrokapseln verpackte Fremdstoffe, ein Problem der Bodenuntersuchung. Stand und Leistung agrikulturchemischer und agrarbiologischer Forschung XXII. Sonderheft zur Zeitschrift "Landwirtschaftliche Forschung", 249–252.

Stephen, I. 1960. Clay orientation in soils. Science Progress 48:322–331.

Stolt, M. H., and D. Lindbo. 2010. Soil organic matter. Pages 129–148 in G. Stoops, V. Marcelino, and F. Mees, editors. Interpretation of micromorphological features of soils and regoliths. Elsevier, Amsterdam, The Netherlands.

Stoops, G. 1964. Application of some pedological methods to the analysis of termite mounds. In: A. Bouillon, editor, Etude sur les termites africains. Eiditions de l'Universite,i Leiopoldville. Universite,i Leiopoldville, South Africa. p. 379–398.

Stoops, G. 1977. Kwalitatieve en kwantitatieve aspekten van de bodem-mikromorfologie. Natuurwet. Tijdschr. 59:6–20.

Stoops, G. 1978a. Provisional notes on micropedology. R.U.G., International Training Centre for Post-Graduate Soil Scientists, Gent, Belgium.

Stoops, G. 1978b. Some considerations on quantitative soil micromorphology. In: M. Delgado, editor, Micromorfologia de suelos. Proc. 5th Intern. Working Meeting Soil Microm., Granada. p. 1367–1384.

Stoops, G. 1981. Visual aid for the estimation of grain sizes in thin sections. An. Edafol. Agrobiol. 40:2289–2291.

Stoops, G. 1994. Soil thin section description: Higher levels of classification of microfabrics as a tool for interpretations. In: A.J. Ringrose-Voase and G.S. Humphreys, editors, Soil micromorphology: Studies in management and genesis. Developments in Soil Science 22. Elsevier Science Publishers, Amsterdam. p. 317–325.

Stoops, G. 1998. Key to the ISSS: Handbook for soil thin section description. Natuurwet. Tijdschr. (Ghent) 78:193–203.

Stoops, G. 2003. Guidelines for analysis and description of soil and regolith thin sections. Soil Science Society of America, Madison, WI.

Stoops, G. 2008. Micromorphology. In: W. Chesworth, editor, Encyclopedia of soil science. Springer, Amsterdam. p. 458–466.

Stoops, G. 2009a. Seventy years "micropedology" 1938-2008: The past and future. Journal of Mountain Science 6:101–106. doi:10.1007/s11629-009-1025-3

Stoops, G. 2009b. Evaluation of Kubiena's contribution to micropedology: At the occasion of the Seventieth Anniversary of his book "micropedology". Eurasian Soil Sci. 42:693–698. doi:10.1134/S1064229309060155

Stoops, G. 2013. A micromorphological evaluation of pedogenesis on Isla Santa Cruz (Galápagos). Spanish Journal of Soil Science 3:14–37.

Stoops, G. 2014. The "fabric" of soil micromorphological research in the 20th century- A bibliometric analysis. Geoderma 2013:193–202. doi:10.1016/j.geoderma.2013.08.017

Stoops, G. 2015a. Análisis de la contextura de la masa basal y los rasgos edáficos del suelo. In: J.C. Loaiza, G. Stoops, R. Poch, and M. Casamitjana, editors, Manual de micromorfología de suelos y técnicas complementarias. Fondo Editorial Pascual Bravo, Medellin, Colombia. p. 87–154.

Stoops, G. 2015b. Composition de la massa basal y de los edaforasgos. In: J.C. Loaiza, G. Stoops, R. Poch, and M. Casamitjana, editors, Manual de micromorfología de suelos y técnicas complementarias. Fondo Editorial Pascual Bravo, Medellin, Colombia. p. 155–204.

Stoops, G. 2017. Fluorescence microscopy. In: C. Nicosia and G. Stoops, editors, Archaeological soil and sediment micromorphology. John Wiley & Sons, Ltd, Chichester. p. 393–397. doi:10.1002/9781118941065.ch36

Stoops, G. 2018. Micromorphology as a tool in soil and regolith studies. In: G. Stoops, V. Marcelino, and F. Mees, editors, Interpretation of micromorphological features of soils and Regoliths. Elsevier, Amsterdam. p. 1–19. doi:10.1016/B978-0-444-63522-8.00001-2

Stoops, G., and H. Eswaran, editors. 1986. Soil micromorphology. Van Nostrand Reinhold Company, New York.

Stoops, G., and A. Jongerius. 1975. Proposal for a micromorphological classification of soil materials. I. A classification of the related distributions of fine and coarse particles. Geoderma 13:189–199. doi:10.1016/0016-7061(75)90017-8

Stoops, G., and A. Jongerius. 1977. Proposals for a micromorphological classification of soil materials. I. A classification of related distribution of coarse and fine particles. A reply. Geoderma 19:247–249. doi:10.1016/0016-7061(77)90032-5

Stoops, G., and F. Mees. 2018. Groundmass: Composition and fabric. In: G. Stoops, V. Marcelino, and F. Mees, editors, Interpretation of micromorphological features of soils and regoliths. 2nd ed. Elsevier, Amsterdam. p. 73–125. doi:10.1016/B978-0-444-63522-8.00005-X

Stoops, G., and C. Nicosia. 2017. Sampling for soil micromorphology. In: C. Nicosia and G. Stoops, editors, Archaeological soil and sediment micromorphology. John Wiley & Sons, Ltd, Chichester. p. 383–391. doi:10.1002/9781118941065.ch35

Stoops, G., and C.E.G.R. Schaefer. 2018. Pedoplasmation: Formation of soil material. In: G. Stoops, V. Marcelino, and F. Mees, editors, Interpretation of micromorphological features of soils and regoliths. 2nd ed. Elsevier, Amsterdam. p. 59–71. doi:10.1016/B978-0-444-63522-8.00004-8

Stoops, G., and T. Tursina. 1992. New methodology for soil thin section description. Pochvovedenye 5:117–121.

Stoops, G., H.-J. Altemüller, E.B.A. Bisdom, J. Delvigne, V.V. Dobrovolsky, E.A. Fitzpatrick, G. Paneque, and J. Sleeman. 1979. Guidelines for the description of mineral alterations in soil micromorphology. Pedologie (Gent) 29:121–135.

Stoops, G., V. Marcelino, S. Zauyah, and A. Maas. 1994. Micromorphology of soils of the humid tropics. In: A.J. Ringrose-Voase and G.S. Humphreys, editors, Soil micromorphology: Studies in management and genesis. Developments in Soil Science. Vol. 22. Elsevier, Amsterdam. p. 1–15.

Stoops, G., E. Van Ranst, and K. Verbeek. 2001. Pedology of soils within the spray zone of the Victoria Falls (Zimbabwe). Catena 46:63–83. doi:10.1016/S0341-8162(01)00153-9

Stoops, G., V. Marcelino, and F. Mees, editors. 2010. Interpretation of micromorphological features of soils and regoliths. Elsevier, Amsterdam.

Stoops, G., M.C. Canti, and S. Kapur. 2017a. Calcareous mortars, plasters and floors. In: C. Nicosia and G. Stoops, editors, Archaeological soil and sediment micromorphology. John Wiley & Sons, Ltd, Chichester. p. 189–199. doi:10.1002/9781118941065.ch23

Stoops, G., A. Tsatskin, and M.C. Canti. 2017b. Gypsic mortars and plasters. In: C. Nicosia and G. Stoops, editors, Archaeological soil and sediment micromorphology. John Wiley & Sons, Ltd, Chichester. p. 201–204. doi:10.1002/9781118941065.ch24

Stoops, G., S. Sedov, and S. Shoba. 2018a. Regoliths and soils on volcanic ash. In: G. Stoops, V. Marcelino, and F. Mees, editors, Interpretation of micromorphological features of soils and regoliths. 2nd ed. Elsevier, Amsterdam. p. 721–751. doi:10.1016/B978-0-444-63522-8.00025-5

Stoops, G., V. Marcelino, & F. Mees. 2018b. Micromorphological features and their relation to processes and classification. General guidelines and overview. In: G. Stoops, V. Marcelino, and F. Mees, editors, Interpretation of micromorphological features of soils and regoliths. Second Edition. Elsevier, Amsterdam, p. 895-917.

Stoops, G., V. Marcelino, & F. Mees, editors. 2018c. Interpretation of micromorphological features of soils and regoliths. Second Edition. Elsevier, Amsterdam.

Tardy, Y. 1993. Pétrologie des Latérites et des Sols Tropicaux. Masson, Paris.

Tippkötter, R. 1990. Staining of soil microorganisms and related materials with fluochromes. In: L.A. Douglas, editor, Soil micromorphology: A basic and applied science. Elsevier, Amsterdam. p. 605–611.

Tröger, W.E. 1969. Optische Bestimmung der gesteinbildenden Minerale. Teil 2. Textband. Schweizerbart'sche Verlagsbuchhandl., Stuttgart.

Tröger, W.E. 1971. Optische Bestimmung der gesteinbildenden Minerale. Teil 1. Bestimmungstabellen. Schweizerbart'sche Verlagsbuchhandl., Stuttgart.

Tucker, M.E. 2001. Sedimentary petrology. An introduction to the origin of sedimentary Rocks. Third ed. Blackwell Science, Oxford.

Tudhope, A.W., and M.J. Risk. 1985. Rate of dissolution of carbonate sediments by microboring organisms, Davies Reef, Australia. J. Sediment. Petrol. 55:440–447.

Turc, G., J.-M. Triat, S. Sassi, H. Paquet and G. Millot. 1985. Caractères généraux de l'épigénie carbonatée de surface, par altération météorique liée à la pédogenèse et par altération sous couverture liée à la diagenèse. C.R. Acad. Sc. Paris, 300, Série II, 7: 283–290.

Twiss, P.C., E. Suess, and R.M. Smith. 1969. Morphological classification of grass phytoliths. Soil Sci. Soc. Am. Proc. 33:109–115. doi:10.2136/sssaj1969.03615995003300010030x

Upton, R., A. Graff, G. Joliffe, R. Länger, and E. Williamson. 2011. American herbal pharmacopoeia: Botanical pharmacognosy- microscopic characterization of botanical medicines. American Herbal Pharmacopoeia, CRC Press, Boca Raton, FL.

Van Dam, D., and L.J. Pons. 1972. Micromorphological observations on pyrite and its pedological reaction products. Acid Sulphate Soils. Proc. Intern. Symposium, Wageningen, 169–195.

Vanden Bygaart, A.J., and R. Protz. 1999. The representative elementary area (REA) in studies of quantitative soil micromorphology. Geoderma 89:333–346. doi:10.1016/S0016-7061(98)00089-5

van der Meer, J.J.M. 1996. Micromorphology. In: J. Menzies, editor, Past glacial environments. Sediments, forms and techniques. Vol. 2. Butterworth-Heinemann, Oxford. p. 335–355.

van der Meer, J.J.M. 1997. Particle and aggregate mobility in till: Microscopic evidence of subglacial processes. Quat. Sci. Rev. 16:827–831. doi:10.1016/S0277-3791(97)00052-8

van der Meer, J.J.M., and J. Menzies. 2011. The micromorphology of unconsolitated sediment. Sediment. Geol. 238:213–232. doi:10.1016/j.sedgeo.2011.04.013

Van Vliet-Lanoë, B. 1980. Approche des conditions physico-chimiques favorisant l'autofluorescence des minéraux argileux. Pedologie (Gent) 30:369–390.

Van Vliet-Lanoë, B., and C.A. Fox. 2018. Frost action. In: G. Stoops, V. Marcelino, and F. Mees, editors, Interpretation of micromorphological features of soils and regoliths. 2nd ed. Elsevier, Amsterdam. p. 575–603. doi:10.1016/B978-0-444-63522-8.00020-6

Van Vliet-Lanoë, B., J.-P. Coutard, and A. Pissart. 1984. Structures caused by repeated freezing and thawing in various loamy sediments: A comparison of active, fossil and experimental data. Earth Surf. Process. Landf. 9:553–565. doi:10.1002/esp.3290090609

Veneman, P.L.M., M.J. Vepraskas, and J. Bouma. 1976. The physical significance of mottling in a Wisconsin toposequence. Geoderma 15:103–118. doi:10.1016/0016-7061(76)90081-1

Vepraskas, M. 1992. Redoximorphic features for identifying aquic conditions. North Carolina Agric. Res. Serv. Techn. Bul. 301. North Carolina Agric. Res. Serv., Raleigh, NC. p. 33.

Vepraskas, M.J., A.G. Jongmans, M.T. Hoover, and J. Bouma. 1991. Hydraulic conductivity of saprolite as determined by channels and porous groundmass. Soil Sci. Soc. Am. J. 55:932–938. doi:10.2136/sssaj1991.03615995005500040006x

Vepraskas, M.J., L.P. Wilding, and L.R. Drees. 1994. Aquic conditions for soil taxonomy: Concepts, soil morphology and micromorphology. In: A.J. Ringrose-Voase and G.S. Humphreys, editors, Soil micromorphology: Studies in management and genesis, Developments in Soil Science 22. Elsevier Science Publishers, Amsterdam. p. 117–131.

Vepraskas, M.J., D.L. Lindbo, and M.H. Stolt. 2018. Redoximorphic features. In: G. Stoops, V. Marcelino, and F. Mees, editors, Interpretation of micromorphological features of soils and regoliths. 2nd ed. Elsevier, Amsterdam. p. 425–445. doi:10.1016/B978-0-444-63522-8.00015-2

Vergouwen, L. 1981. Scanning electron microscopy applied on saline soils of the Konya Basin in Turkey and from Kenya. In: E.B.A. Bisdom, editor, Submicroscopy of soils and weathered rocks. Pudoc, Wageningen. p. 237–248.

Verleyen, E., K. Sabbe, and W. Vyverman. 2017. Siliceous microfossils from single-cell organisms: Diatoms and chrysophycean stomatocysts. In: C. Nicosia and G. Stoops, editors, Archaeological soil and sediment micromorphology. John Wiley & Sons, Ltd, Chichester. p. 165–170. doi:10.1002/9781118941065.ch19

Verrecchia, E.P. 1990. Litho-diagenetic implications of the calcium oxalate biogeochemical cycle in semiarid calcretes, Nazareth, Israel. Geomicrobiol. J. 8:87–99. doi:10.1080/01490459009377882

Verrecchia, E.P., P. Freytet, K.E. Verrecchia, and J.-L. Dumont. 1995. Spherulites in calcrete laminar crust: Biogenic $CaCO_3$ precipitation as a major contributor to crust formation. J. Sediment. Res. A65:690–700.

Villagran, X., and R.M. Poch. 2014. A new form of needle-fiber calcite produced by physical weathering of shells. Geoderma 213:173–177. doi:10.1016/j.geoderma.2013.08.015

Villagran, X.S., D.J. Huisman, S.M. Mentzer, C.E. Miller, and M.M. Jans. 2017. Bone and other skeletal tissues. In: C. Nicosia, and G. Stoops, editors, Archaeological soil and sediment micromorphology. John Wiley & Sons Ltd, Chichester. p. 11–38. doi:10.1002/9781118941065.ch1

Vogel, H.-J. 1994. Mikromorphologische Untersuchungen von Anschliff-Präparaten zur räumlichen Porengeometrie in Böden im Hinblick auf Transportprozesse. Ph. D. Diss. Univ. Hohenheim, Stuttgart.

Vrydaghs, L. 2017. Opal sponge spicules. In: C. Nicosia and G. Stoops, editors, Archaeological Soil and Sediment Micromorphology. John Wiley & Sons, Ltd, Chichester. p. 171–172. doi:10.1002/9781118941065.ch20

Vrydaghs, L., and Y. Devos. 2018. Visibility, preservation and colour: A descriptive system for the study of opal phytoliths in (Archaeological) soil and sediment thin sections. Environmental Archaeology doi:10.1080/14614103.2018.1501867

Vrydaghs, L., Y. Devos, and A. Petö. 2017. Opal phytoliths. In: C. Nicosia and G. Stoops, editors, Archaeological soil and sediment micromorphology. John Wiley & Sons, Ltd, Chichester. p. 155–163. doi:10.1002/9781118941065.ch18

Warne, S.St.J. 1962. A quick field laboratory staining scheme for the differentiation of the major carbonate minerals. J. Sediment. Petrol. 32:29–38.

Weibel, E. 1979. Stereological methods. Vol. 1. Practical methods for biological morphometry. Academic Press, London.

Werner, J. 1962. Uber die Hertsellung fluoreszierender Bodenanschliffen. Z. Pflanzenernlihr. Diing. Bodenkd. 99:144--150.

West, L.T., S.C. Chiang, and L.D. Norton. 1992. The morphology of surface crusts. In: M.E. Summer and B.A. Stewart, editors, Soil crusting: Chemical and physical processes. Lewis Publ., Boca Raton. p. 73–92.

Wieder, M., and D.H. Yaalon. 1974. Effect of matrix composition of carbonate nodule crystallisation. Geoderma 11:95–121. doi:10.1016/0016-7061(74)90010-X

Wilding, L.P. 1997. Micromorphology emphasis in the United States over the past decade. p. 13-22. In S. Shoba et al. (ed.) Soil micromorphology. Studies on soil diversity, diagnostics, dynamics. Agricultural Univ. of Wageningen, the Netherlands.

Wilding, L.P., and L.R. Drees. 1968. Distribution and implication of sponge spicules in surficial deposits in Ohio. Ohio J. Sci. 68:92–99.

Wilding, L.P., and L.R. Drees. 1990. Removal of carbonates from thin sections for microfabric interpretation. In: L.A. Douglas, editor, Soil micromorphology: A basic and applied science. Elsevier, Amsterdam. p. 613–620. doi:10.1016/S0166-2481(08)70377-5

Wilding, L.P., and K.W. Flach. 1985. Micropedology and soil taxonomy. In: L.A. Douglas and M.L. Thompson, editors, Soil micromorphology and soil classification. SSSA Special Publication Nr. 15. SSSA, Madison, WI. p. 1–16.

Williams, A.J., M. Pagliai, and G. Stoops. 2018. Physical and biological surface crusts and seals. In: G. Stoops, V. Marcelino, and F. Mees, editors, Interpretation of micromorphological features of soils and regoliths. 2nd ed. Elsevier, Amsterdam. p. 539–574. doi:10.1016/B978-0-444-63522-8.00019-X

Winchell, A.N. 1962. Elements of optical mineralogy. Part Ill, Determinative tables. 2nd Ed. John Wiley & Sons. New York.

Winchell, A.N., and H. Winchell. 1951. Elements of optical mineralogy. Part II. Descriptions of minerals. John Wiley & Sons, New York. doi:10.1080/11035895109453350

Yanguas, J.E., and J.J. Dravis. 1985. Blue fluorescent dye technique for recognition of microporosity in sedimentary rocks. J. Sediment. Petrol. 55:600–602. doi:10.2110/jsr.600

Yardley, B.W.D., W.S. MacKenzie, and C. Guilford. 1990. Atlas of metamorphic rocks and their textures. Longman, Essex, UK.

Zaniewski, K. 2001. Plasmic fabric analysis of glacial sediments using quantitative image analysis methods and GIS techniques. PhD thesis. University of Amsterdam, Amsterdam, The Netherlands.

Zaniewski, K. & JJM. van der Meer, 2005. Quantification of plasmic fabric through image analysis 63:109–127.

Zainol, E., and G. Stoops. 1986. Relationship between plasticity and selected physico-chemical and micromorphological properties of some inland soils from Malaysia. Pedologie (Gent) 36:263–275.

Zauyah, S., C.E.G.R. Schaefer, and F.N.B. Simas. 2018. Saprolites. In: G. Stoops, V. Marcelino, and F. Mees, editors, Interpretation of micromorphological features of soils and regoliths. 2nd ed. Elsevier, Amsterdam. p. 37–57. doi:10.1016/B978-0-444-63522-8.00003-6

Zimmerle, W. 1991. Thin-section petrography of pelites, a promising approach in sedimentology. Geol. Mijnb. 70:163–174.

Zingg, T.H. 1935. Beitrag zur Schotteranalyse. Schweiz. Mineral. Petrogr. Mitt.15:39–140.

APPENDIX
Materials, Light, and the Petrographic Microscope

A.1 INTRODUCTION

The optical properties of nonopaque materials will be briefly reviewed here to introduce readers to basic concepts and terminology of optical mineralogy. This appendix contains only those data necessary for understanding the optical effects observed in a thin section with a polarizing or petrographic microscope. More advanced methods, such as the determination of axial figures, are not discussed in this Appendix. For more details readers are referred to excellent manuals such as Parfenoff et al. (1970), Phillips (1971), Phillips and Griffen (1981), Gay (1982), Pichler and Schmitt-Riegraf (1997), Puhan (1994), Nesse (2003) and Raith et al. (2011).

A.2 ANISOTROPIC AND ISOTROPIC MATERIALS

Materials can be subdivided into crystalline and noncrystalline (amorphous or short range order) materials. Crystalline materials have atoms arranged in a rigid three-dimensional pattern called a *crystal lattice*. Most minerals are crystalline and *anisotropic* which means that the values for a single physical characteristic (e.g., speed of light transmission, hardness) will change depending upon the orientation of the crystal. In noncrystalline materials, the atoms show a random arrangement and they behave as *isotropic* bodies. Values for properties such as the speed of light remain the same regardless of the direction that the light moves through the material. Examples of isotopic materials are gasses, liquids, and glass. In thin sections, the most important isotropic components are the glass slide and cover glass, the resin used to impregnate and mount the sample, volcanic glass, and gels. Some minerals that have been called *amorphous* have in fact a short range order and behave as optical isotropic bodies. Examples in soils are minerals such as allophane, ferrihydrate,

Guidelines for Analysis and Description of Soil and Regolith Thin Sections, Second Edition. Georges Stoops.
© 2021 Soil Science Society of America, Inc. Published 2021 by John Wiley & Sons, Inc.
doi:10.2136/guidelinesforanalysis2

and opal,. Crystals belonging to the cubic crystal system (e.g., garnet, halite, diamond) are also optical isotropic. Isotropic bodies may exhibit anisotropic characteristics under stress. This can be for instance the case in the resin used for impregnation (Plate 9.1d and e).

With respect to light transmission, materials can be subdivided into *transparent* and *opaque*, although this may be function of the thickness and of the wavelength of the light. Nontransparent minerals can become transparent when thinned. In petrography, minerals are considered as opaque when no light passes at standard thickness of a thin section (i.e., 20 to 30 µm). Many Fe-, Mn-, and Ti-oxide minerals fit into this category.

A.3 PROPERTIES OF LIGHT

For our purposes light can be considered simply as a continuous spectrum of electromagnetic radiation in the form of waves. The color of a light ray is determined by its *wavelength* λ (the distance between two identical points on a wave). The visible spectrum is situated between wavelengths of 380 nm (transition from ultraviolet to violet) and 780 nm (transition from red to infrared). White light contains a mixture of rays having all these wavelengths and is termed polychromatic light.

Interference of two waves having the same wavelength (Fig. A1) can either result in a more intense wave, providing the interfering waves are in phase (waves differ a whole number of wavelengths), or to an annihilation of the waves if their phase difference is 1/2 λ, 1½ λ, 2½ λ, etc.

In normal light, the waves vibrate in all planes perpendicular to the ray's axis of propagation. If the vibration of a light wave is restricted to a single plane, it is called *polarized light*. Normal light can be polarized for instance by passing through a *polarizer*, a sheet of polaroid for example.

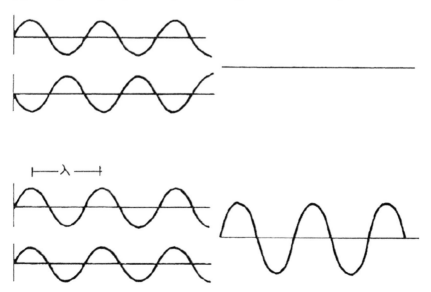

Fig. A1. Interference of two rays of a same wavelength, but with a phase difference. When the phase difference corresponds to a whole number of wavelengths (nλ), the intensity will increase. When the difference is 1/2 λ, 1½ λ, 2½ λ, annihilation will take place.

A.4 LIGHT AND OPTICAL ANISOTROPIC BODIES

When white light passes through a transparent body, some wavelengths may be selectively absorbed, and a colored object is observed. In the case of optical anisotropic minerals, absorption can be different for different directions. This means that, when plane polarized light is used, a mineral grain may show different colors when rotated. This phenomenon is called *pleochroism*. It is strongly expressed in minerals such as biotite, tourmaline, hornblende and vivianite, but may be visible also in some Fe-stained clay coatings.

When a light ray enters a transparent, optical isotropic body the ray is refracted. For example, a light beam traversing a glass plate is refracted when entering the glass, and again when leaving to enter the air. When entering an optical anistropic body, however, the light is *double refracted*: it is split into two beams which are polarized in perpendicular planes. Each beam has its own speed of propagation (one called *fast*, corresponding with the lowest refractive index, and the other *slow*, corresponding with the highest refractive index) and hence, a different refractive index for a given orientation of the crystal. The numerical value of the difference between the highest (slow ray) and the lowest (fast ray) refractive index is called the *birefringence*, and is an important characteristic in mineral identification.

A.5 OBSERVATIONS WITH THE PETROGRAPHIC MICROSCOPE

A.5.1 The petrographic microscope

The petrographic or polarizing microscope (see Fig. A2) is a microscope adapted to studies of minerals in polarized light. Its principal features will be traced below.

The light emitted by the light source (e.g. an halogen bulb), is polarized by the *polarizer* which is build-in below the condenser. It is transmitted through the condenser as a parallel beam (orthoscopic observation). An additional lens, the front lens, which can be inserted above the condenser, transforms this to a light cone, allowing observations in conoscopic light in order to determine the axial figure of the mineral (not discussed in this introduction), but also used to increase the light intensity at high magnifications.

The thin section or grain mount is clamped on a circular graduated rotating stage.

The removable *analyser* in the microscope tube above the objective is identical to the polarizer, but turned over 90° with regard to the latter, which means that all the light transmitted by the polarizer will be adsorbed in the analyser when the latter is slid into the microscope tube (so called *crossed polarizers*); no light come through.

Between the sample and the analyzer a *retardation plate* or *compensator* can be introduced. This is a metal frame with an oriented mineral plate fitting in a special slot (Fig A2, n° 9) of the microscope, oriented at 45° with respect to the polarizer and analyzer. The orientation of the vibration direction with the largest refractive index nγ (thus the slowest ray) is indicated on this compensator. The most useful for

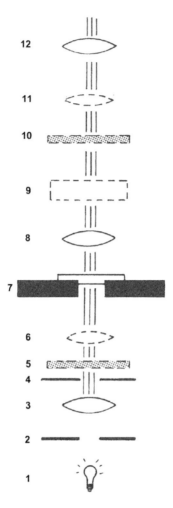

Fig. A.2. Components of a polarizing microscope, from bottom to top: (1) the light source (generally a halogen bulb), (2) the stage diaphragm, (3) the condenser lens, (4) the field diaphragm, (5) the polarizer, generally oriented N-S, (6) the swing out condenser (front lens), (7) the graduated rotatable stage with the thin section, (8) the objective lens, (9) a slot for the compensator, (10) the removable analyzer (i.e. a polarizer with E-W vibration direction), (11) the removable lens of Bertrand, and (12) the eyepiece.

micromorphological studies is the so called gypsum red compensator or 1λ retardation plate. Less popular is the use of the $\frac{1}{4}\lambda$ retardation plate (so called mica plate), that changes the interference colors only $\frac{1}{4}\lambda$ of a wavelength (Plate 4.3c and f).

A.5.2 Observation with polarizer only (Plain Polarized Light, PPL)

In PPL, the following features can be studied: shapes, cleavages, color, including pleochroism, by rotating the stage (Plate 8.13a: compare the color of the horizontally and the vertically oriented vivianite crystals) and *relief* which is a mineral property related to the difference in refractive index between a mineral and its environment. If the difference is very small, the border of the mineral will be visible only with difficulty, and the mineral is said to have a *low relief* (Plate 4.4a); if the difference is large, the mineral will stand out clearly and show clear dark boundaries, it is said to have a *high relief* (Plate 6.1a: compare the low relief of the feldspar at left with the high relief of the hornblende grain, Plate 6.1e).

A.5.3 Observation with polarizer and analyzer (Crossed Polarized Light, XPL).

When an optically isotropic body (e.g., glass, garnet, (Plate 6.3c and d), halite (Plate 8.2c and d)) is placed between the polarizer and the analyzer, no light will reach the eyepiece because the body has not altered the light rays and the vibration direction of polarizer and analyzer are perpendicular to each other.

When a plane polarized ray of light enters an anisotropic body, the ray will be split into two perpendicular component rays that will advance through the mineral with different speeds because the mineral has different refractive indices for different orientations as explained in Section 4 (Fig. A3). Both component rays will exit the crystal with a phase difference. If they enter the analyzer at an oblique angle to its vibration direction, each component ray leaving the mineral will again be split into two component rays. One of these component rays will vibrate parallel to the vibration direction of the analyzer and will be transmitted. The second component ray will vibrate perpendicular to the vibration direction of the analyzer and will be absorbed. The original component rays that passed through the crystal and the analyzer will have and opposite vibration direction and also a phase difference and will interfere. If the phase difference equals a whole number of wavelengths (e.g. 1λ, 2λ, 3λ... $n\lambda$) annihilation will take place. For white light this results in the removal of some colors from the spectrum and changes the color of the light. The combination of different wavelengths of polychromatic light that pass the analyzer produces a phenomenon called *interference color*. Interference color is an optical phenomenon and not an inherent mineral property. It is dependent on mineral orientation, thickness (Plate 9.2f), and birefringence. If the phase difference is small, a dark gray interference color is visible. With gradually larger phase differences, the following interference colors appear: light gray, yellow, red, blue, green, again yellow, red, blue, green, etc. This sequence is repeated and known as colors of first order (gray to red), second order (red to red), third order, etc. (Fig. A4). In each successive order, the colors become fainter until the light appears nearly white (so called *white of higher order*).

When the vibration directions of the mineral under observation are parallel to that of the polarizer or analyzer, no light will pass. The mineral is said to be in an *extinction position* and appears black. By rotating the stage, however, the mineral will gradually illuminate and interference colors will appear. This extinction can be overcome by using so called circular polarized light (see Section 3.2.2.2).

For elongated grains (prismatic, acicular) or elongated grain sections (e.g., section perpendicular to the cleavage of a mica) it is possible to

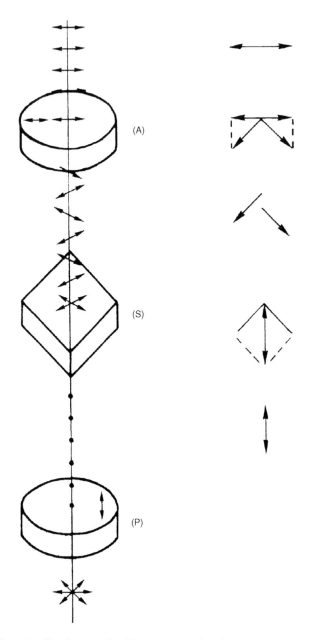

Fig. A.3. The formation of interference colors. The ray of normal light (bottom) is polarized when passing through the polarizer (P). When entering the sample (S) it is split in two rays with perpendicular vibration directions and different speed. In the analyser (A) both rays interfere and are recombined into a single one. Each time the phase difference corresponds to a whole number of wavelengths annihilation takes place. In the case of white light this results in the appearance of the complementary colors.

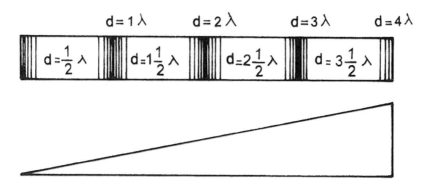

Fig. A4. A quartz wedge observed between crossed polarizers shows interference colors of several orders. Colors of the first order (d = 1 λ) are grey, yellow, orange and red; those of the second order purple, blue, green, yellow, orange and red (d = 2 λ), etc..

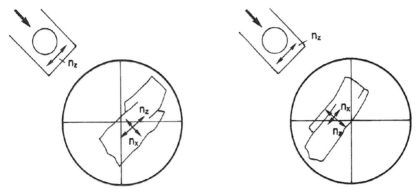

Fig. A5. Use of the λ retardation plate (compensator, gypsum plate) to determine the orientation of the fast and the slow ray in an elongated mineral. (x: lowest refractive index, fastest ray, z: highest refractive index, slowest ray).

determine whether the *fast* or the *slow* vibration direction corresponds to the elongated direction, by inserting a compensator or retardation plate. If the slow ray of the retardation plate correspond to the slow ray of the mineral section, the interference colors increase to colors of a higher order, and the mineral or feature is said to be *length slow*. If the slow ray of the retardation plate correspond to the fast ray of the mineral section, the interference colors decrease to colors of a lower order, the mineral or feature is said to be *length fast*. One has to bring the elongated grain or feature (e.g., a clay coating) parallel to the vibration direction of the slowest ray of the compensator (generally perpendicular to the frame of the compensator) (Fig. A5, Plate 4.3b and c, Plate 6.10b and c, e and f, Plate 8.6b and c, e and f), and observe whether the interference colors have gone up or down one order of the scale for the 1λ compensator (gypsum plate), or ¼ order for the ¼ λ compensator (mica plate) (Plate 4.3d and e) .

Subject Index

Bold: definition or explanation

italic: figure, plate or table

A

Abundance 42, **57**, *59*, 73, 90, 154, 195, 196
Accommodation
    aggregates 77, **82**
    voids 71, **75**, 77
Acicular 60, 61, **62**, 111, 161, 229
Ageing of excrements 178
Aggregate nodule 171, **171**, *173*
Aggregates 68, *74*, 80, 81, 86
Aggregation 67, 68, **77**, 79
Air bubbles 27, 35, 71, *132*, *140*, *147*, *191*, *192*, *193*
Alteration
    degree of 92, 195
    general 63, **90**, 115, 144, 157
    inorganic components 104, *109*, 111, 113
    mineral *32*, 90, 91, *98*, 99, 100, 102, 168
    organic material 121, *124*, *125*, 178
    patterns *89*, 93, **94**, *96*, *97*, *96*
    rock 98
Alteromorph **94**, 100, 105, *171*, 172, 194
Alteromorphic nodule *171*, **172**, *175*
Aluminium (fluorescence) 27
Amorphous organic fine material **125**, *122*, 125, *132*, 160, 180
Analyzer 21, *134*, 137, 141, 227, *228*, 229
Anhedral 91, 93, 174
Animal residues **120**

B

Anisotropic 46, 47, 131, 133, 141, 174, 225, 227, 229
Angular *62*, 88, 104, *136*, *145*, 172, 191, 196
    aggregates or peds *73*, *74*, *76*, **77**, *78*, 80, *81*, 82
    microstructure 76, 78, 83, **85**
Anorthic nodule *31*, *134*, **169**, *171*, *187*
Anthropogenic elements 88, **113**, 115
Apedal 67, 68, 77, *79*, 85
Arrangement 6, 38, 39, *42*, 47, 52, *55*, 67, 73, 82, 107, 114, 130, 138, 148, 163, 165, *171*, 190
Artifacts 10, **191**

B

Banded
    alteration pattern *32*, 94
    distribution 43, **43**, 44, 47, 82, 114, *122*, 130, *139*, 157
Basaltic glass *89*, *101*
Basaltoid *101*
Basic
    constituents 9, 40, 52, 63, 69, 85, 88, 194
    distribution patterns *43*, **43**, 54, 82, 181, 197
    microstructure 83, **83**, 195
    orientation patterns *42*, *44*, **45**, 45, 55, 131, 154, 181

Guidelines for Analysis and Description of Soil and Regolith Thin Sections, Second Edition. Georges Stoops.
© 2021 Soil Science Society of America, Inc. Published 2021 by John Wiley & Sons, Inc.
doi:10.2136/guidelinesforanalysis2

Printed and bound by CPI Group (UK) Ltd, Croydon, CR0 4YY
09/09/2021
03082273-0001